艺海无涯。
我刚刚启航,
无论碧波万顷还是雨骤风狂,
我深信,
艺术的航船
都会驶入一个美丽的港湾。

SAILING IN THE OCEAN OF ART

CHINA FORESTRY PUBLISHING HOUSE

艺海启航

刘传刚盆景艺术轨迹

◎ 刘传刚 著
Liu Chuangang

中国林业出版社

藝海啟航

刘传铭签名东艺术航迹

阁泽苏

序 言
Foreword

WBFF 世界盆景友好联盟主席胡运骅
President, World Bonsai Friendship Federation

中国盆景界人才济济，技艺高超者众多，但能进入高等院校系统地讲授盆景课程的屈指可数，能登上国际盆景讲坛并获得成功者更如凤毛麟角，而刘传刚先生两者皆能。

传刚艺高胆大，曾赴新加坡、马来西亚和美国作过盆景演讲和创作示范。今年4月又应西班牙盆景协会之邀与赵庆泉、黄就伟一起去西班牙参加表演，与日本、印度尼西亚以及欧洲的盆景专家同场竞技。他创作的风动式榆树盆景和风动式博兰水旱盆景造型新颖别致，特式鲜明，给欧洲盆景爱好者留下深刻的印象。场内三百多位观众都有耳目一新的感觉。表演结束后掌声经久不息，观众纷纷上台与刘传刚合影留念，意大利盆景协会随即发出邀请，希望他明年9月赴意大利再作表演。

传刚师承艺术理论功底深厚的中国盆景艺术大师贺淦荪教授，年轻时即打下了扎实的艺术基础。十多年前到海南艰苦创业，物华天宝、人杰地灵的海南使他如鱼得水。传刚和当地的盆景爱好者充分开发利用当地博兰、香兰等优秀的盆景素材，创作出一大批盆景佳作，使海南盆景在全国盆景大展中屡获金奖，传刚成为海南盆景的领军人物。

传刚大胆探索，敢于创新。尽管我对他正在探索的"雨林式"盆景的名称，树石盆景的分类等有不同的看法，但他这种探索、创新的精神我十分赞赏。

《艺海启航——刘传刚盆景艺术轨迹》，是一部综合性的盆景著作，内容丰富、全面、图文并茂，既有创作技法的讲解，又有艺术理论的阐述，且树木盆景、山石盆景、水旱盆景……都有涉及。深信《艺海启航》的出版一定深受广大读者的欢迎！

China has abundance of capable people in the field of penjing. However, Mr. Liu Chuangang stands out among the talents since he has not only delivered lectures on penjing in the institutions of higher learning, but also has successful experience in many international forums.

Mr. Liu has given lectures and demonstration of penjing creation in Singapore and America. Invited by the Spanish penjing association this year, Liu went to Spain with Mr. Zhao Qingquan and Mr. Huang Jiuwei in April to show their skills, competing with bonsai experts from Japan, Indonesia and Europe. His works of dynamic elm penjing and Poilaniella fragilis landscape penjing impressed the European audiences by novelty and unique features. His show evoked continued applause and invitation from Italy for next year.

Mr. Liu was the apprentice of Professor He Gansun, a famous penjing master in China with a solid theoretical foundation. More than a decade ago, Mr. Liu came to Hainan to carve out his career under arduous conditions, and finally achieved great success. He created a pile of penjing by local trees like Poilaniella fragilis and Chinese cymbidium, and became a leading light in Hainan after receiving golden prizes in national exhibitions repeatedly.

Mr. Liu is bold in blazing new trails. Although I have different opinions on the name he gave to the "Rainforest Penjing" and his classification of the Tree-and-Rock Penjing, I still appreciate his spirit of exploration and innovation.

"Sailing in the Ocean of Art", Liu Chuangang penjing art track, is a comprehensive book with introduction to both creating skills and artistic theory, which is excellent in both illustrations and texts. I do believe it will be very popular after publication.

Hu Yunhua

2011 年 5 月 28 日

赏
Appreciation of "Liu Chuangang's Penjing"

中国盆景艺术大师、中国风景园林学会花卉盆景赏石分会顾问

Chinese Penjing Artist, Consultant of Penjing & Shangshi Branch of Chinese Society of Landscape Architecture

近段时间，我一直在思考，我们盆景人应该拿什么来激励自己、要求自己，作为成功的秘诀或成功的标准？最后认为是"德才兼备"、"文武全才"。这八个字听来虽属老生常谈，似缺新意，但仔细品读却很有深意。然不易成功。

前些日子，刘传刚先生寄来《艺海启航——刘传刚盆景艺术轨迹》一书编写提纲，并要求为它写个"点评"或"评语"以示鼓励和鞭策。盛情不可却，于是我想将以上所想提供给刘传刚先生与之共勉。

刘先生师承我国著名的盆景艺术大师贺淦荪教授，风风雨雨三十年，终于在海南打定江山。在事业上，坚持了老师动势盆景的理论精髓，还充分体现了创新精神的雨林式盆景及对新树种——博兰的开发、利用……

作品风格雄浑壮丽、自然清新。作品如人，能给人以格外的注意力和吸引力。

刘先生吃苦耐劳、坚持创新、能做能写、力求全面发展的创业精神，令人鼓舞。

Recently, I've been thinking about the essential factors of success for the penjing people like us. The conclusion I summed up is to be graced with both virtues and talents, as well as to be excellent in intellectual pursuits and practical skills. However, it is easier said than done.

I received the outline of "Liu Chuangang's Penjing" the other day, and was invited to make comments on it. As a result, I presented my views, and hopefully could make joint efforts with Mr. Liu.

As an apprentice of the famous penjing artist, Professor He Gansun, Mr. Liu has been ahead of his career in Hainan after 30 years of hardships. Not only adhered to develop dynamic penjing, what he had learned from Mr. He, Liu also devoted himself to create new styles, like rainforest penjing and Poilaniella fragilis Penjing.

Liu's works can always attract others' attention with vigorous and natural features.

Mr. Liu is hard-working and bold in blazing new trails. Now he is about to publish a book on the basis of his practical experiences, which is very inspiring.

Hu Yueguo
2011 年 5 月 29 日

观
Views on "Liu Chuangang's Penjing"

中国盆景艺术大师、中国风景园林学会花卉盆景赏石分会副理事长

Chinese Penjing Artist, Vice Chairman of Penjing & Shangshi Branch of Chinese Society of Landscape Architecture

在我的印象中，刘传刚先生是一位创新意识极强的盆景艺术家。他青年时期师从中国盆景艺术大师贺淦荪先生，接受了"飞扬动势"的理念和"树石盆景"的手法。来到海南后又大胆创新，勇于实践，利用当地"天然大温室"的气候特点和资源优势，创出了具有"动势"特色和"树石"、"水旱"形式的博兰盆景系列。其作品雄浑流畅，个性鲜明，具有很强的视觉冲击力。

According to my impression, Liu Chuangang is a penjing artist who constantly brings forth new ideas in the arts. He apprenticed with He Gansun in his young age, a famous Chinese penjing artist, learning the idea of "Dynamic Penjing" and the method of producing "Tree-and-Rock Penjing". Bold in blazing new trails, he creates a series of penjing (Poilaniella fragilis) on the basis of Hainan's climate and resources, including dynamic style, tree-and-rock style and water-landscape style. His works are flowing and vigorous, with specific characters and visual effect.

Zhao Qingquan
2011 年 6 月 3 日

全国政协副主席、台盟中央主席林文漪（右），参观刘传刚锦园并题字

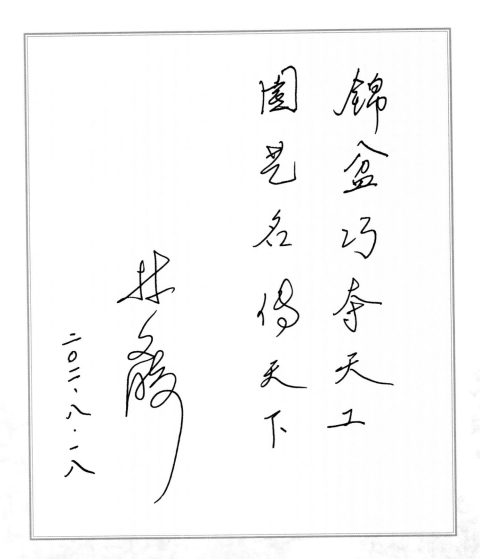

锦盆巧夺天工
园艺名传天下

林文漪

二〇一〇·八·八

全国政协人口资源环境委员会副主任委员、中共海南省委原书记汪啸风（左）题字

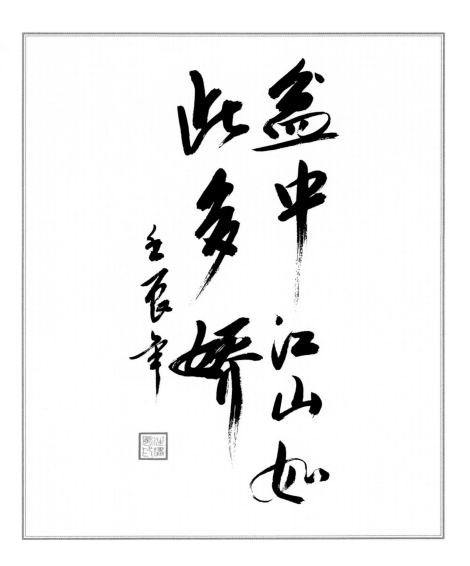

海南省政协副主席、中国书法家协会理事王应际（右）题字

北京大学中国画法研究院院长、
当代著名学者、国学大师、教授
范曾题字

Inscriptions ◎题字

中国盆景艺术家协会永久名誉会长、《中国花卉盆景》杂志社总顾问苏本一题字

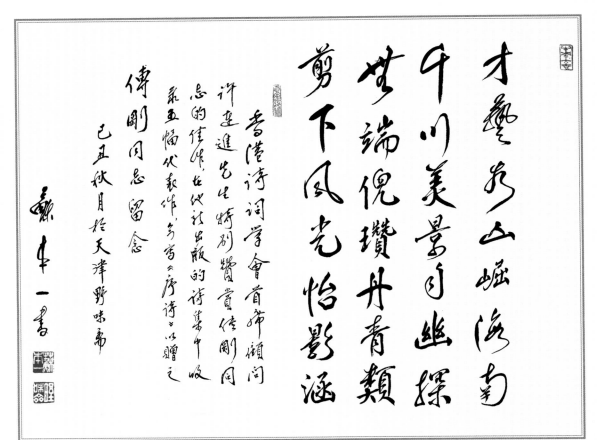

才艺为山崛海南
千川美景自幽深
丝端倪瓒丹青颜
剪下风光怡影涵

香港诗词学会首席顾问许达进先生特别赞赏傅剛同志的作品，在他社出版的诗集中收录五幅代表作，今写《序诗》以赠之

傅剛同志留念

己丑秋月於天津野味斋

苏本一书

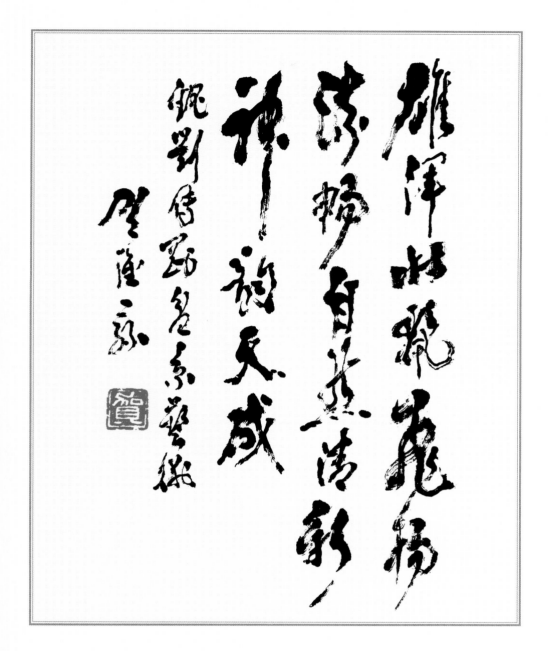

中国盆景艺术大师、教授贺淦荪（右）为弟子刘传刚《艺海启航》一书题字

目录

序言·············胡运骅 / 6
评语·············胡乐国 / 8
　　　　　　赵庆泉 / 9
题字·············林文漪 / 10
　　　　　　汪啸风 / 11
　　　　　　王应际 / 12
　　　　　　范　曾 / 13
　　　　　　苏本一 / 14
　　　　　　贺淦荪 / 15

■ 创作探索篇

◎ 第一章　动势盆景创作探索　/ 22

第一节　动势盆景的理论探索　22
　一、动势盆景的由来　22
　二、动势盆景的理论基础　22
　三、动势盆景的主要形式　24
　四、风动式树木盆景的制作　26
第二节　动势盆景的作品展示　28
第三节　动势盆景的创作演示　36
　一、动势盆景创作演示之一　36
　二、动势盆景创作演示之二　38
　三、动势盆景创作演示之三　40
　四、动势盆景创作演示之四　42
　五、动势盆景制作要领　43

◎ 第二章　雨林式盆景创作探索　/ 44

第一节　雨林式盆景的理论探索　44
　一、雨林式盆景的由来　44
　二、雨林式盆景基本形式　45
　三、雨林式盆景的创作难度　45
　四、雨林式盆景的制作　46
第二节　雨林式盆景的作品展示　48
第三节　雨林式盆景的创作演示　56
　一、雨林式盆景创作演示之一　56
　二、雨林式盆景创作演示之二　58
　三、雨林式盆景制作要领　59

◎ 第三章　树木盆景创作探索　/ 60

第一节　树木盆景的主要形式　60
　一、直立式树木盆景的造型　60
　二、倾斜式树木盆景的造型　61
　三、悬崖式树木盆景的造型　62
　四、丛林式树木盆景的造型　63
第二节　树木盆景的作品展示　66
第三节　树木盆景的创作演示　78
　一、树木盆景创作演示之一　78
　二、树木盆景创作演示之二　80
　三、树木盆景创作演示之三　82
　四、树木盆景制作要领　83

◎ 第四章　山石盆景创作探索　/ 84

第一节　山石盆景的主要形式与分析　84
　一、峰状山石盆景造型　85
　二、岩状山石盆景造型　85
　三、岭状山石盆景造型　85
　四、石状山石盆景造型　86
　五、组合山石盆景造型　86
第二节　山石盆景的作品展示　88
第三节　山石盆景的创作演示　102
　一、山石盆景创作演示之一　102
　二、山石盆景创作演示之二　104
　三、山石盆景制作要领　105

◎ 第五章　树石盆景创作探索　/ 106

　第一节　树石盆景主要创作形式探讨　106
　　　一、点石式　106
　　　二、倚石式　107
　　　三、坡岭式　107
　　　四、环抱式　108
　　　五、临崖式　108
　　　六、高峰式　109
　第二节　树石盆景的作品展示　110
　第三节　树石盆景的创作演示　120
　　　一、树石盆景创作演示之一　120
　　　二、树石盆景创作演示之二　122
　　　三、树石盆景创作演示之三　124
　　　四、树石盆景制作要领　125

◎ 第六章　小型与微型盆景创作探索　/ 126

　第一节　小型与微型盆景主要创作形式探讨　126
　　　一、盆景的规格尺寸　126
　　　二、小型与微型盆景的由来　126
　　　三、小型与微型盆景的用材、造型与管理　127
　　　四、小型与微型盆景的陈设　129
　第二节　小型和微型盆景的作品展示　130
　第三节　小型和微型盆景的创作演示　138
　　　一、小型盆景创作演示　138
　　　二、微型盆景创作演示　140
　　　三、小型和微型盆景制作要领　141

▍艺术交流篇

◎ 第七章　国际交流　/ 144
　　　一、欧洲"三国行"纪实　144
　　　二、世界盆景大会——"美洲行"纪实　146
　　　三、日本国风展——"日本行"纪实　148
　　　四、马来西亚盆景古石展赛会纪实　150
　　　五、菲律宾第七届亚太盆景会议纪实　151

◎ 第八章　国内交流　/ 152
　　　一、主持海南省"三届"盆景评比展览　152
　　　二、台湾盆景艺术活动　154
　　　三、主持"世界盆景友好联盟交流中心"落户海南活动　155
　　　四、香港岭南盆景艺术活动　156
　　　五、广州国际盆景邀请展　157
　　　六、上海"东沃杯"中国盆景精品展暨创作比赛　158
　　　七、广东（陈村）BCI国际盆景赏石博览会　159
　　　八、北京第八届亚太地区盆景赏石会议暨展览会　160
　　　九、刘传刚盆景个性化邮票首发式　161

◎ 第九章　海南花卉大世界建设　/ 162
　　　一、海南花卉大世界简介　162
　　　二、海南花卉大世界创建情况　163
　　　三、海南花卉大世界倍受政府关怀　163
　　　四、海南花卉大世界系列活动　164

刘传刚盆景艺术轨迹　/ 168

后记　/ 170

Contents

春意盎然（六月雪、海母石）
Furious Spring(Serissa Japonica,Corallite)

Foreword / 6

Appreciation of "Liu Chuangang's Penjing" / 8
Views on "Liu Chuangang's Penjing" / 9

Inscriptions / 10

Theory Exploration

I A Survey of Dynamic Penjing / 22
Theory Exploration of Dynamic Penjing /22
The origin of Dynamic Penjing /22
Theoretical basis of Dynamic Penjing /22
Major forms of Dynamic Penjing /24
Producing of windswept penjing style /26
Works of Dynamic Penjing /28
Creation demonstration of Dynamic Penjing /36
Creation demonstration of Dynamic Penjing(1) /36
Creation demonstration of Dynamic Penjing(2) /38
Creation demonstration of Dynamic Penjing(3) /40
Creation demonstration of Dynamic Penjing (4) /42

II A Survey of Rainforest Penjing / 44
Theory Exploration of Rainforest Penjing /44
The origin of Rainforest Penjing /44
Main Styling Forms of Rainforest Penjing /44
Difficulties in the creation of Rainforest Penjing /45
Training of Rainforest Penjing /46
Works of Rainforest Penjing /48
Creation demonstration of Rainforest Penjing /56
Demonstration of Rainforest Penjing(1) /56
Demonstration of Rainforest Penjing(2) /58

III A Survey of Tree Penjing / 60
Main Styling Forms of Tree Penjing /60
Training of upright penjing style /60
Training of slanting penjing style /61
Training of hanging cliff penjing style /62
Training of forest penjing style /63
Works of Tree Penjing /66
Creation demonstration of Tree Penjing /78
Demonstration of Tree Penjing(1) /78
Demonstration of Tree Penjing(2) /80
Demonstration of Tree Penjing(3) /82

IV A Survey of Landscape Penjing / 84
   Main Styling Forms and Analysis of Landscape Penjing /84
   Peak-shape Landscape Penjing /85
   Cliff-shape Landscape Penjing /85
   Range-shape Landscape Penjing /85
   Stone-shape Landscape Penjing /86
   Combination Landscape Penjing 86
   Works of Landscape Penjing /88
   Creation demonstration of Landscape Penjing /102
   Demonstration of Landscape Penjing(1) /102
   Demonstration of Landscape Penjing(2) /104

V A Survey of Tree-Rock Penjing / 106
   Main Styling Forms of Tree-Rock Penjing /106
   Ornament Style /106
   Leaning Style /107
   Range Style /107
   Circle Style /108
   Cliff Style /108
   Peak Style /109
   Works of Tree-Rock Penjing /110
   Creation demonstration of Tree-Rock Penjing /120
   Demonstration of Tree-Rock Penjing(1) /120
   Demonstration of Tree-Rock Penjing(2) /122
   Demonstration of Tree-Rock Penjing(3) /124

VI A Survey of Miniature Penjing / 126
   Main Styling Forms of Small and Miniature Penjing /126
   Sizes of penjing /126
   The origin of small and miniature penjing /126
   Materials, shaping and maintaining of small and miniature penjing /127
   Display of small and miniature penjing 129
   Works of Tree-and-Rock Penjing /130
   Creation demonstration of Tree-and-Rock Penjing /138

▌ Arts Exchange

VII International Exchange / 144
   · Trip in Europe /144
   · The World Bonsai Convention /146
   · The Kokufu-bonsai Exhibition in Japan /148
   · The International Penjing & Gushi Exhibition & Contest in Malaysia /150
   · The Asia-Pacific Penjing & Shangshi Exhibition & Convention in the Philippines /151

VIII Domestic Exchange / 152
   · Hainan Penjing Exhibition /152
   · Art activities in Taiwan /154
   · Exchange centre of World Bonsai Friendship Federation in Hainan /155
   · The Membership Representative Conference of Hong Kong Lingnan Penjing Institute /156
   · Guangzhou International Penjing Invitational Exhibition /157
   · The Dow Cup China Penjing Invitation Exhibition & Penjing Creation Contest /158
   · China (Chencun) International Penjing & Shangshi Expo /159
   · The 8th Asia Pacific Penjing & Shangshi Exhibition & Convention /160
   · Release of "Personal Stamp Sheet of Liu Chuangang Penjing" /161

IX Construction of Hainan Flower World / 162

Personal Record of Liu Chuangang /168

Postscript /170

九天揽月（博兰）
Pluck Moon out of Sky(Poilaniella fragilis)

创作探索篇
Theory Exploration

- 第一章　动势盆景创作探索　A Survey of Dynamic Penjing
- 第二章　雨林式盆景创作探索　A Survey of Rainforest Penjing
- 第三章　树木盆景创作探索　A Survey of Tree Penjing
- 第四章　山石盆景创作探索　A Survey of landscape Penjing
- 第五章　树石盆景创作探索　A Survey of Tree-Rock Penjing
- 第六章　小型和微型盆景创作探索　A Survey of Miniature Penjing

第一章
动势盆景创作探索
A Survey of Dynamic Penjing

第一节
动势盆景的理论探索
Theory Exploration of Dynamic Penjing

一、动势盆景的由来
The origin of Dynamic Penjing

"动"是由一种姿势或形态转变为另一种姿势或形态;"势"是指姿势或形态改变的速度或力度。例如我们在散步、跑步、跨越等运动时,只要在动,就由一种姿势转变成另一种姿势,而转变速度的快慢和力度的强弱,就会产生出不同的势,这就是我们常常说的动势(图1-1)。在盆景的艺术造型中,如风中的树木,由于风力的作用,造成了树木由一种形态改变成另一种形态,将这种形态通过树木造型表现出来,这就是动势盆景中的风动式盆景造型(图1-2)。

The meaning of "dynamic" here is two-fold: to alter a position or a shape, and the speed and intensity of the alteration. FFor example, walking, running, jumping, etc. all imply a dynamic state caused by different speed and intensity (Figure 1-1). In the design of miniature trees, there is a typical form of Dynamic Penjing called windswept style, which represents the alteration of tree branches due to wind force (Figure 1-2).

二、动势盆景的理论基础
Theoretical basis of Dynamic Penjing

宇宙万物都处在生生不息的运动之中,运动是绝对的,静止是相对的;变化是绝对的,统一是相对的。正如恩格斯说的:"绝对的静止,无条件的平衡,是不存在的,个别运动趋向平衡,可总的运动又破坏平衡。"

Everything in the universe is in constant motion and eternally changing. Motion is absolute while stagnation is relative. The philosophical basis of Dynamic Penjing is just as what Friedrich Engels said: "There is no such thing as absolute rest, unconditional equilibrium. Each separate movement strives towards equilibrium, and the motion as a whole puts an end again to the equilibrium."

1. 动和静 Motion and rest

在艺术形式美的法则中,动与静既对立又统一,既相互对抗又相互依赖。没有动,静不能存在,没有静也就无所谓动。在动势盆景的造型中,强调"动"在造型

图1-1 春洒人间(三角梅;曲干、半悬崖动势造型)
Figure 1-1, Spring Reigns Everywhere (Bougainvillea, slanted trunk, hanging cliff penjing style)

图1-2 大风歌（博兰、海母石；大型组合风动式造型）
Figure 1-2, Song of Wind (Poilaniella fragilis, corallite, windswept style)

中的主导地位，以造成生动、活泼、动荡、险峻、激昂甚至于紧张的气氛。而要在乱中求整，险中求稳，则需要用"静"去调节和缓解"动"造成的不安和紊乱，以达到和谐统一的目的。

In the beauty of artistic form, motion and rest are the unity of opposites. They are contradictory as well as interdependent. The concept of "motion", which works for a vivid, vigorous, undulating, precipitous, roused or even tense effect, occupies the leading position in the training and shaping of Dynamic Penjing. Nevertheless, well-organized and steady shaping skills, what "rest" is related to, are indispensible to balance the unrest elements caused by "motion", aiming at perfect harmony.

2. 平衡与均衡 Equality and balance

以中线将形体分成两边，左右对称的形体称为平衡。平衡属于"静"的造型元素，它整齐，庄重、严肃，同时也带有一些呆板的味道。在造型中，体积大小虽然有别，在视觉上感到重量相等而达到的"平衡"称为之均衡。例如在天平上，一边是棉花，一边是铁块，重量相等却大小有别，因为能"衡"，所以既有变化，又能统一，既有"动"感，又达到了稳定，这就是均衡。"秤砣虽小压千斤" 也是说的均衡。

Equality means bilateral symmetry divided by a central line. It is a shaping element out of "rest" that leads to regular, solemn, serious while inflexible result. On the other hand, speaking of balance in styling penjing, it means the "equality" that causes a visual effect of equal weight despite of different volume and sizes. For example, an iron block and cotton of equal weight that lay on each side of scales are different in sizes. However, balance of alteration and unity, motion and rest is kept in this case.

3. 动势和平衡 Motion and even

在造型艺术中，动势和平衡，两者矛盾的变化达到统一就是要力求达到均衡。一方面在造型上力求变化以"动"造成千姿百态，另一方面要动中有静，以克服动极而不安的矛盾，以求得造型中和谐统一。平衡趋入静，导致造型上的平稳往往显得平淡，如果要求得生动，就需要在静中求动，如此反复交替，周而复始。动势和平衡相互制约又相互依存，在这种交替中去升华，可以创造出多种形式的美感。这就是动势盆景的美学原理。

In the art of shaping, contradiction of motion and even can be transformed to reach unification by balance. For one thing "motion" produces all kinds of shapes and postures, and for anther "rest" is required to relax the excessive tension. Being even tends to rest, and that may lead to steady but flat works. The level of harmony will be attained by the appropriate combination of these two elements. The alternation of motion and even, which are contradictory while interdependent, produces diversified aesthetic perception. That is the aesthetic principle of Dynamic Penjing.

三、动势盆景的主要形式
Major forms of Dynamic Penjing

树木在大自然的生长过程中，从幼苗到大树，经历着各种不同的从一种形态转变成另一种形态，所以说"动"是千姿百态的。而且不同的树种，千差万别的生长空间更是造成千变万化的条件，造就了千姿百态的树木形象。就动势盆景的造型而论，上扬、倾斜的枝干是动；横倚、下垂的枝干还是在动。因此风吹、雪压能造成树木的动势；拟人手法的漫舒广袖、编钟乐舞的树木造型更具有旋律感的动势；还有那行云流水的飘逸，金凤腾飞的扶摇长天、风驰电掣的长驱直入，如图1-3至图1-10。动势盆景的造型要求"创作无定型、规律有共性"，依据创作立意的主题，挖掘与创造能表情达意的各种动势造型，表现各种不同的动势。因此，动势盆景造型不搞固定的模式，遵循"从有法到无法"，"从无我到有我"，不拘一格地发展个性，以达到"随心所欲不越矩"的境界。但是，在动势盆景中，以风动式的树木造型的动势最为强烈，最为典型，对风动式树木造型的领悟、理解和把握，对其他的动势树木的造型将起到提纲挈领的作用。（插图绘制：唐吉青）

图1-3 疑是银河落九天（深圳盆景世界收藏）（博兰）
Falling Galaxy(Collected By Shenzhen Bonsai World)(Poilaniella fragilis)

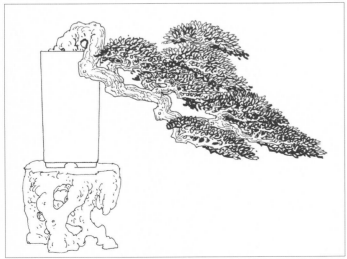

图1-4 悬崖式 Hanging cliff penjing style

图1-5 垂枝式 Weeping style

From young to old, trees keep altering shapes during their growing process in nature. What is more, trees of all kinds grow into various forms since conditions vary from place to place. Motion in the shaping of Dynamic Penjing concerns upward, downward, slanting and transverse branches. From Figures 1-3 to 1-10, some representative works are shown: shaping like blown by wind and bowed by snow, styling with natural and unrestrained branches, and training to rise steeply and grow straightly. In the shaping of Dynamic Penjing, it is required to produce works of particularity while following rules of generality, and to create kinds of dynamic postures according to the artistic theme. The windswept style is the most typical Dynamic Penjing, which causes the strongest feeling of motion. Consequently, learning and understanding the windswept style is crucial for the shaping of Dynamic Penjing.

图 1-7 奔击式 Thrusting style

图 1-6 腾跃（台湾郑国瑞先生收藏）（博兰）
Making Way (Collected By Zheng Guorui, Taiwan)(Poilaniella fragilis)

图 1-8 压雪式 Snow-load style

图 1-9 临水式 Facing water style

图 1-10 风动式 Windswept style

四、风动式树木盆景的制作
Producing of windswept penjing style

1. 按意选材 Material selection on artistic conception

盆景创作立意为先,特别是风动式盆景创作更应该如此。树木因受到风的压力,树的枝干转变成风的运动方向,才由一种姿势转变成另一种姿势。风力有强弱大小之分,气候季节不同,树木枝干老幼粗细不一,对枝干造型时弯曲的程度是不一样的。应该强调作品创作的主题,需要表现怎样的情境:是清风徐徐还是狂风怒吼,萧瑟的秋风还是和煦的春风,对于树木枝干种种不同的细节处理,要细致认真地推敲和把握。

按意选材,是要把符合创作主题思想的独具个性的材料挑选出来,并且还不能忽视树木桩坯材料的许多共性要求:如树木整体构架完美、根盘完整、根理健壮、树干走向顺畅、结构合理、收尾自然、出枝茂密……(图1-11)。

Conception is the first step of penjing creation, especially for producing the windswept style. The branches as well as the trunk grow to one side as though the wind has been blowing the tree constantly in one direction. The bending degree of branches varies in line with different wind intensity, seasons and the ages and thickness of the tree. As a result, every detail of branches and trunk should be properly dealt with after deliberation.

Material selection should focus on the ideas of guiding artistic creation and be representative. However, there are some common selecting standards, such as a tree with integral structure, complete and strong roots, and branches with smooth trends, suitable structure, natural tip ends, luxuriant twigs and so on(Figure 1-11).

2. 见机取势 Determined courses according to specific conditions

风动式盆景的造型,势取向背乃是树木造型"布势"的关键所在。分枝布于四方:向势力求长枝,强化动势;背势蓄枝转向,前枝遮掩,后枝陪衬,一弛一张是动势造型的精要。盆景蓄枝不可能尽如人意,如有残缺,借枝转向,以达到整体动态的完整。作者盆景造型的智慧主要体现在"见机取势"这四个字的含义之中,见机:需要灵活,更需要积累和修养;取势:要合目的性,达到造型的要求和原则。

To decide branches go with or against the wind plays a pivotal role in training windswept penjing style. The branches grow out on all sides of the trunk but will all eventually be bent to one side. It is best for the longer branches to go in the following direction and strengthen the dynamic feeling. Besides, the other branches can be retained to change direction: some are interspersed in front and some serve as a contrast behind. Combination of tension and relaxation is the essence of Dynamic Penjing. Nevertheless, branch retaining will not be entirely satisfactory for the most of time, so more branches need to change direction to perfect the works. That is why a qualified maker should be

图 1-11 造型前审视全貌,确定正、反、向、背面(Figure 1-11)

flexible and artistically accomplished to create works in conformity with the shaping requirements.

3. 骨架枝的定位 Positioning of frame branches

风动式树木造型的骨架枝的定位,要在正常树木骨架的基础上,突出向背弛张,造型上要奇中求正、险中求稳、动中有静,克服和打破左右平衡的形态,定向、定点、定芽蓄枝,创造良好的生长环境,任其放长,并适当控制顶和上部分的枝条,放纵下部分特别是第一级枝的生长。骨架枝的定位要下疏上密,聚散合理,整体的定位要比一般树木造型的定位疏朗些,适当留出空间以便将来造型。

The positioning of frame branches for windswept penjing style is to highlight the directions and density of branches. It is necessary to combine uniqueness and regularity, motion and rest appropriately, and break the bilateral symmetry. On the other hand, directional branches ought to be retained from chosen buds in certain places. With favorable growing environments, branches are allowed to attain required length. Generally speaking, the top and higher branches need to be controlled, whereas the lower ones are given free rein to grow, especially those of first rank. All in all, branches should meet and part with reasonable density, and be thinner and clearer as a whole than other general works, so there could be enough space for later shaping.

4. 风动式造型的蟠扎 Wiring skills of windswept penjing style

按照风动式树木造型的骨架枝，长到满意的程度，造型前施肥1～2次，让枝条健壮以便蟠扎。蟠扎前一至二天不要浇水，摘去全部叶片，清除掉不需要的枝条，先确定主枝托和大层面，由下至上，先扎顺向枝，再扎逆向的回头枝，依顺序操作。造型定位时要注意拉开枝托与枝托之间的距离，要留出各层次未来的发展空间，为后来的枝条生长留有余地，否则，死板一块，容易造成挫枝和缺片（图1-12）。

风动式树木顺势和枝条，下部分一、二级枝托，分枝平垂，略向下倾斜，结顶和上部分的枝托弯曲倾斜要呈现上扬的形态，中部的枝托介于上下部分之间的过渡。

最能表现风吹树木的枝条是反向的"抗力枝"，就是逆向的回头枝，它是最强的受风点。"抗力枝"的造型，不能全部向上翻，否则会造成"背驼驼"，堵死枝条生长的空间，造成死枝挫片，它的造型要注意左、右弧线的运用。"抗力枝"在风动式树木的造型中最重要、最具观赏价值，但又是最难长好的，故要认真培养，必要时先放长枝条，健壮后再造型处理（图1-12）。

图1-12 使用金属丝先固定好枝条基部再顺时针绕开芽点捆绑 (Figure 1-12)

风动式树木造型结顶的位置，是调剂树木的重心，求得均衡的重要因素。直立式和倾斜式的树木风动造型中，在结顶处引垂线至树根部，朝向势略为前倾为好。这与人跑步时从头部引垂线，人的头部前于脚部的道理是一致的，但结顶的位置过分向前，容易与下部的枝条齐平，错落不开，弄不好成为直角三角形，破坏动感，对动势造型产生不利的影响。只要了解结顶在树木造型中能起协调重心的作用，那么在树木造型中，就会正确处理结顶的位置了。

制作风动势树木盆景，需要毅力和恒心，每年都需要不断地蟠扎、解丝，每根枝条都需要多次造型，随着枝干的粗壮和增多，造型的难度也愈来愈大，作品也随之愈来愈精致，在造型过程中，享受创作的挑战和考验的乐趣！

When the frame branches of windswept penjing have grown to satisfactory length, fertilization should be done for 1 to 2 times before shaping, so that the branches would be strong enough for wiring. Watering is not allowed 1 or 2 days before wiring, and all the leaves and unnecessary branches need to be cleared up. The process of wiring must be carried out in proper order. First of all, we should determine the principal branches and layers. Then from bottom to top, branches that go in the following direction ought to be wired before those need to be reversed. Furthermore, we have to increase the distance between each main branch for further development. Otherwise, it may result in damaged branches and incomplete shape(Figure 1-12).

Among the branches that go in the following direction, the lower ones should be laid horizontally and incline downward slightly, while the top and higher ones should be bent upward. Those in the middle can be seen as the transitional area.

The "resistant branches", which encounter the strongest wind force, are reversed to show the constant blowing. They can not be upturned entirely. Otherwise there will not be enough room for growing. That is why we should make good use of curved lines. The "resistant branches" are always the most important and eye-catching aspects of a windswept tree. Nevertheless, it is not easy for them to grow well. They require careful training, and sometimes should not be bent until they are long and strong enough(Figure 1-12).

The position of the tree's apex is a crucial element to attain balance. It is better to lean towards the following direction for both upright and slanting styles of windswept penjing, just in the same way of a running man, whose head always precedes the foot. However, if the apex is over drawn, it may easily fall on a line with the lower branches, forming a right triangle, and result in disordered branches and no moving feeling. That is to say, the position of an apex can balance the shape as a whole.

It requires tenacity and persistence to create windswept penjing, because the wires need to be wrapped and removed every year, and each branch may be retrained for many times. As the branches increase and grow, it will be more and more difficult to produce delicate works. However, that is also the fun of creation.

第二节
动势盆景的作品展示
Works of Dynamic Penjing

◎ 作品名称：宝岛雄风
◎ 树种：博兰
◎ 石种：灰石
◎ 树高：80cm
◎ 盆长：120cm
◎ 心得：作品选用一棵卧干（三干）博兰树进行造型，充分利用树木的"卧"势，并将全部枝条从左向右反方向蓄枝，动中求静、险中求稳，达到均衡效果。主树干基部右下角及远景处靠接两组枝条与主树相呼应，丰富画面，增强意境。缺点是用盆稍偏小，水面空间太少。

◎ **Title of Work:** Windswept Island
◎ **Tree species:** Poilaniella fragilis
◎ **Stone kind:** Limestone
◎ **Tree height:** 80cm
◎ **Basin length:** 120cm
◎ **Creation experience:** A lying trunk style (three trunks) Poilaniella fragilis is used for shaping. Making the best of the tree's lying posture, all the branches are trained from the left to the right, being opposite to their original direction. The branches on the lower right-hand corner and the back of the lying trunk add brilliance to the work's picturesque scene. However, the basin is a little bit small so that there is not enough space for the water surface.

- ◎ 作品名称：风雷激
- ◎ 树种：博兰
- ◎ 石种：灰石
- ◎ 树高：90cm
- ◎ 盆长：130cm
- ◎ 心得：选用一株直干博兰树进行造型，主干垂直，但全部枝条由东向西飞舞，似东风劲吹；配石也由东向西蜿蜒流动，树石呼应，相得益彰。缺点是树木主干太直，枝条层次有待梳理。

- ◎ **Title of Work:** Hit by wind & thunder
- ◎ **Tree species:** Poilaniella fragilis
- ◎ **Stone kind:** Limestone
- ◎ **Tree height:** 90cm
- ◎ **Basin length:** 130cm
- ◎ **Creation experience:** An upright-trunk Poilaniella fragilis is used, while all the branches are trained from the east to the west, just like blown by easterly wind. The chosen stones are also winding in the same direction. The tree and stones bing out the best in each other. Nevertheless, the trunk is too straight, and the branches could be arranged more appropriately.

- ◎ 作品名称：天涯劲风
- ◎ 树种：博兰
- ◎ 石种：英石
- ◎ 树高：98cm
- ◎ 盆长：150cm
- ◎ 心得：三株树木表现近、中、远景，基干（下半部）直立，中间主干和上部树干弯曲较大，以显其劲风之力度。作品以线条的变化突出"飞扬的动势"，并达到"写意的效果"。缺点是用盆稍小了一些，左边土层可适当加高点，右边一棵树木可适当降低。

- ◎ **Title of Work:** Strong wind from the remotest corner of the earth
- ◎ **Tree species:** Poilaniella fragilis
- ◎ **Stone kind:** Limestone from Yingde
- ◎ **Tree height:** 98cm
- ◎ **Basin length:** 150cm
- ◎ **Creation experience:** Thees three trees show different perspectives to the viewer. The lower half of the trees is straight while the upper is bent obviously in order to show the strength of the wind, manifesting the dynamic feeling. However, the basin is comparatively small. On the other hand, if the soil layer on the left is higher, and the tree on the right is lower, so much the better.

- 作品名称：飞天梦
- 树种：榆树
- 石种：火山石
- 树高：70cm
- 盆长：72cm
- 心得：这是一件中型风动式树石盆景，此株风动榆树造型20多年，枝条舒展奔放，犹如书法之"狂草"。作品造型三曲有向，其势欲飞，有遨游太空之意。缺点是部分枝条稍长了些。

- Title of Work: Flying dream
- Tree species: Ulmus pumila
- Stone kind: Volcanic rock
- Tree height: 70cm
- Basin length: 72cm
- Creation experience: It is a medium-size windswept tree-rock penjing which has been trained for over 20 years. All the branches are extending like a wild scribble with twists and turns, whereas some of them are too long.

- ◎ 作品名称：风满楼
- ◎ 树种：博兰
- ◎ 石种：海母石
- ◎ 树高：69cm
- ◎ 盆长：68cm
- ◎ 心得：树木取势均衡，虽山雨欲来，狂风怒吼，但树木悬根露爪、紧依盆面，不可动摇。该作品人工之痕甚少，宛若天成。缺点是树木右下角处应有一块适当的点石，重心会更稳。

- ◎ **Title of Work:** Atmosphere of Wind
- ◎ **Tree species:** Poilaniella fragilis
- ◎ **Stone kind:** Corallite
- ◎ **Tree height:** 69cm
- ◎ **Basin length:** 68cm
- ◎ **Creation experience:** The tree manifests a state of balance. It stands still steadily under a furious storm. The work is quite natural as an integral whole. Nevertheless, it may be better if one big stone was added on the low right corner.

- ◎ 作品名称：东风劲吹
- ◎ 树种：对节白蜡
- ◎ 石种：龟纹石
- ◎ 树高：70cm
- ◎ 盆长：72cm
- ◎ 心得：树木取势向背，层次活泼，线条流畅，盆面起伏，野味十足。缺点是主流枝不突出，右边抗力枝粗度不够。

- ◎ Title of Work: Strong wind from the east
- ◎ Tree species: Fraxinus hupehensis
- ◎ Stone kind: Weathered granite
- ◎ Tree height: 70cm
- ◎ Basin length: 72cm
- ◎ Creation experience: The branches are trained against their original growing direction, and the arrangement of them is quite clear and vivid. However, the major branches have not been highlighted, and the ones on the right are not strong enough.

- 作品名称：海岛风云（意大利 Crespi 盆景博物馆收藏）
- 树种：博兰
- 石种：云石
- 树高：66cm
- 盆长：130cm
- 心得：此件作品是用"景盆法"的方法将一棵中型博兰直接种在一块天然云石上，进行了近 20 年的造型，云石象征海岛，旁边再点缀两小岛作衬景。实显风云变幻、深化主题。缺点是云盆石坡脚略显僵硬了些，若再圆润或变化多一点会更好。

- Title of Work: Stormy Island (Collected by Italian Crespi Bonsai Museum)
- Tree species: Poilaniella fragilis
- Stone kind: Marble
- Tree height: 66cm
- Basin length: 130cm
- Creation experience: The selected tree has been planted directly on a natural marble for almost 20 years. The marble is the "island", accompanied by two smaller islands by the side. The theme of the work is perfectly shown whereas the slope toe of the stone should be more flexible.

- ◎ 作品名称：东风号
- ◎ 树种：博兰
- ◎ 树高：98cm
- ◎ 盆长：42cm
- ◎ 心得：这株直立式博兰树，通过风动式造型后，酷似一战士在战场上吹响冲锋号角的场景，给人一往无前的精神动力。不足之处是风动枝内再有一主流枝（破平立异）就更好了。

- ◎ **Title of Work:** Easterly Bugle
- ◎ **Tree species:** Poilaniella fragilis
- ◎ **Tree height:** 98cm
- ◎ **Basin length:** 42cm
- ◎ **Creation experience:** Trained into the windswept style, the shape of the tree looks like a soldier sounding the bugle call to charge. On the other hand, the work may be better if one major branch was added among those stretching ones.

第三节
动势盆景的创作演示
Creation demonstration of Dynamic Penjing

一、动势盆景创作演示之一
Creation demonstration of Dynamic Penjing(1)

2009年7月13日，笔者和黄就伟、林南先生一行三人应美国西雅图盆景博物馆（Pacine Rin Bonsai Collection）之邀，一同前往该博物馆做了两场盆景创作演示，该馆馆长Darid J.Degroot组织了当地100多名盆景爱好者来现场观摩，笔者边讲、边演、边做，并与观众互动，现场气氛热烈。图为其中一场动势盆景的创作演示过程。

On July 13th, 2009, Mr. Liu Chuangang, Mr. Huang Jiuwei and Mr. Lin Nan were invited to give a demonstration in Pacine Rin Bonsai Collection. More than 100 devotees participated in the activity organized by Darid J. Degroot. A demonstration of dynamic penjing is illustrated in the following content.

1

2

3

4

5

1. 所表演树木为美国当地一株原生态树木。刘传刚表演前作现场设计；
2. 讲解示范表演的程序。曹伟泰先生（左）担任现场翻译；
3. 对树木摘叶，笔者邀请美国爱好者上台协助；
4. 摘叶后的效果；
5. 对树木枝条造型；
6. 造型后效果并开始脱盆；
7. 将树木脱出；
8. 将树木造型后重新种植和布局；
9. 点缀石头；
10. 铺好苔藓；
11. 作品表演完后效果。曹伟泰先生作现场翻译。

Figure1. Given a local tree, Mr. Liu Chuangang sketched an impromptu design;
Figure2. Mr. Liu explained the steps of his demonstration to the audience, and Mr. Cao Weitai interpreted on the spot;
Figure3. Pick the leaves. Local devotees were invited to assist in this step;
Figure4. Tree without leaves;
Figure5. Shape the branches;
Figure6,7. Get the tree out of the original basin;
Figure8. Replant and arrange the tree in a new basin;
Figure9. Earth up the tree and add some stones;
Figure10. Put on the moss;
Figure11. Finished work.

6

7

8

9

10

11

二、动势盆景创作演示之二
Creation demonstration of Dynamic Penjing(2)

应美国亚太园艺公司总裁曹伟泰先生之邀，笔者于 2009 年 7 月 12 日在该盆景交流中心做了一场风动式盆景示范表演。

（一）表演材料：该中心提供的一盆榆树；

（二）表演内容：改作和重新创作；

（三）表演形式：风动式造型。

Invited by Mr. Cao Weitai, the president of American Asia-Pacific Horticulture Company, Mr. Liu Chuangang gave a demonstration of windswept style penjing in the exchange centre on July 12th, 2009.

Materials: ulmus pumila

Object: to reshape the tree

Style: windswept style

1. 所表演的树木为榆树；
2. 对树木摘叶；
3. 摘叶后的效果；
4. 对树木先清枝和修剪；
5. 开始对树木绑扎；
6. 绑扎造型；
7. 绑扎基本完成对局部适当调整；
8. 调整树木角度；
9. 绑扎后树木俯视效果；
10. 脱盆（将树木拔起）；
11. 重新选用盆（为多边型盆）；
12. 移植在新盆上；
13. 上好培植土后安放石头（缀石）；
14. 在盆面铺植苔藓；
15. 继续铺植苔藓，种植小植物；
16. 笔者示范表演后与曹伟泰先生合影，并将作品命名为"风从东方来"。

Figure1. The prepared tree (the obverse of the ulmus pumila);
Figure2. Pick the leaves;
Figure3. Tree without leaves;
Figure4. Prune the branches;
Figure5, 6, 7. Wire the branches;
Figure8. Make partial adjustment;
Figure9. Angle the branches;
Figure10. Vertical view of the wired tree;
Figure11. Get the tree out of the original basin;
Figure12. Take a polygonal basin;
Figure13. Replant the tree to the new basin;
Figure14. Earth up the tree and add some stones;
Figure15. Put on the moss;
Figure16. Add some little plants;
Figure17. Mr. Cao Weitai and Mr. Liu Chuanguang;
The work was named as "Wind from Orient".

10

11

12

13

14

15

16

三、动势盆景创作演示之三
Creation demonstration of Dynamic Penjing(3)

2010年6月5日，应中国风景园林学会花卉盆景赏石分会之邀，笔者在福建泉州参加了首届中国杂木盆景学术研讨会，期间做了一场风动式盆景示范表演。

Invited by Penjing & Shangshi Branch of Chinese Society of Landscape Architecture, Mr. Liu Chuangang participated the 1st Chinese Weed Tree Penjing Seminar and put on a demonstration of windswept penjing.

1. 表演前的原生态树木（博兰）；
2. 表演用盆为椭圆形紫砂盆；
3. 将表演树木放进盆中，审视如何造型；
4. 摘掉全部树叶；
5. 对枝条绑扎造型；
6、7. 每个枝条都进行绑扎（右为助手连佳荣先生）；
8. 绑扎造型基本完成后，赵庆泉大师作讲解和点评；
9. 对每一枝条再作精细处理；
10. 细心铺植苔藓；
11. 重新种植树木后安放山石；
12. 制作完成后的正面效果；
13. 局部效果；
14. 制作完成后的另一面效果。

1

2

3

4

5

6

7

14

8

9

Figure1. The original tree. (Poilaniella fragilis);
Figure2. Plant the tree in the basin;
Figure3. An elliptic boccaro basin for use;
Figure4. Pick up the leaves;
Figure5. Wire the branches;
Figure6, 7. Create a windswept penjing with the help of Mr. Lian Jiarong;
Figure8. Mr. Zhao Qingquan explained and made comments on the wired tree;
Figure9. Shape the branches;
Figure10. Replant the tree and add some stones;
Figure11. Put on the moss;
Figure12. The obverse of the finished work;
Figure13. A close shot;
Figure14. View from the other side.

10

11

13

12

四、动势盆景创作演示之四
Creation demonstration of Dynamic Penjing(4)

2011年10月22日，应中国第五届盆景学术研讨会组委会之邀，笔者在浙江宁波绿野山庄做了一场盆景创作示范表演。

（一）表演材料：当地提供的两株三角枫、两株小叶黄杨、一株小真柏和一些鸡骨石等。基本属于原生态（未造型）材料。

（二）表演用盆为椭圆形大理石盆，长120cm。

（三）表演形式为风动式树石造型。

Invited by the organizing committee of the 5th Chinese Penjing Academic Forum, Mr. Liu Chuangang put on a demonstration of Penjing creation in Ningbo on October 22nd, 2011.

Materials: trident maple, sinica var microphylla, Sabina chinensis, chicken-bone stones, etc.

Basin: an elliptic marmoreal basin with a length of 120cm

Style: windswept style tree-rock penjing

1. 表演用材和盆；
2. 对树木摘叶；
3. 摘叶后绑扎枝条；
4、5、6. 将造型好树木安放在大理石盆中、细心观察效果；
7. 对树木进行搭配，反复推敲，认真审视，最后培土定位；
8. 制作完成后的效果。

Figure1. Selected materials and basin;
Figure2. Pick the leaves ;
Figure3. Wire the branches;
Figure4, 5, 6. Place trees in the basin;
Figure7. Arrange the materials with deliberation, then earth up the trees;
Figure8. Finished work.

7

8

五、动势盆景制作要领

宇宙万物运动中，动势变化各不同。
神韵自然师造化，大千世界纳入胸。
中得心源溶人情，节奏活泼夺天工。
势取向背定大局，向长背短取势通。
等腰直角是大忌，左右均衡莫居中。
轻歌曼舞腾金凤，垂柳依依沐春风。
雪压风吹势澎湃，飞扬动势写意功。

第二章
雨林式盆景创作探索
A Survey of Rainforest Penjing

第一节
雨林式盆景的理论探索
Theory Exploration of Rainforest Penjing

一、雨林式盆景的由来
The origin of Rainforest Penjing

在整个地球上，没有哪块地方像雨林那么丰富多彩，将人置身于绿色的帷幕高入云霄树林的迷宫之中。几乎每天都有骤雨带来富有营养物质的雨水，树木的生长给人一种压倒一切的优势，强烈程度超过任何其他的森林。雨林中树冠、枝叶稠密，即使中午时分也是光线暗淡，树枝竭力地往高处生长。在这样一个竞争的环境中，各种植物以获得少量阳光求得生存。这是一个永远没有冬天的地方，在这片苍翠的地域中生长的十万多种植物，都必须适应这令人难以置信的复杂生存条件。我曾多次在雨林之中考察，并为这瑰丽而又奇特的景观所陶醉，并产生了用盆景艺术来表现雨林的创作欲望，这是一个艰苦的探索过程，因为在盆景创作领域中没有任何有关雨林的创作经验和方法可以借鉴，只能依靠自己去摸索、研究和实践。

Rainforest is a wonderful ever green world. Growing in a forest with heavy annual rainfall, trees are excessively thick with leaves and always trying to climb up to a higher place. Carrying on the intense competition in a forest without winter, millions of plants and animals living there must adapt to the complex conditions. After times of investigation, my inspiration and enthusiasm for creation of Rainforest Penjing was stimulated. However, it has been a painstaking process since there is

图 2-1 峥嵘（博兰、海母石；雨林式造型）
Figure 2-1 Shooting High (Poilaniella fragilis, corallite)

no previous experience for reference. I am still making repeated attempts on my own way. The following parts are what I sum up about Rainforest Penjing. Please oblige me with your valuable comments.

二、雨林式盆景基本形式
Main Styling Forms of Rainforest Penjing

雨林式盆景理论基础属于动势盆景体系，动是千姿百态的，千差万别的生长空间造成千变万化的树木形态。上扬、倾斜的枝干是动，横倚、下垂的枝干还是在动，

图 2-2 扬帆启航（博兰、云盆、海母石；雨林式造型）Figure 2-2 Setting Sail (Poilaniella fragilis, corallite)

雨林中的树木为争夺生存空间和阳光力求向上生长，突出一个"争"字，"争"是雨林式盆景的一大特色，由于生存竞争，枝干形成向上的直线，以最快捷的速度达到争夺阳光的目的，气根以直线最快地垂落，抢到土壤中的水分来生长。因此，雨林给我最深刻的印象就是一个以直线为主的竞争的世界（图 2-1、图 2-2）。

The theoretical foundation of Rainforest Penjing falls into the system of Dynamic Penjing. Trees in the rainforest are all competing for space and sunshine, trying best to grow higher. Accordingly, "competitiveness" is one of the notable features of Rainforest Penjing. Struggling for survival, branches grow straight upward in order to reach the sunshine fast. On the other hand, the aerial roots drop down vertically, contending for the soil water. As a result, for me, the most impressive image of the rainforest can be described as "a competitive world of straight lines"(Figures 2-1,2-2).

三、雨林式盆景的创作难度
Difficulties in the creation of Rainforest Penjing

（1）由于雨林苍莽密集的气势，那种密度和厚度用盆景的艺术语言的传达比较困难。

（2）地貌复杂，昏暗的底层植物，倒地的腐木、藻类、菌类、苔藓、藤本互相挤压，错综盘结，雨林的地貌环境在小小的盆景之中难以表现。

（3）垂直结构将多数的雨林分为几层，被称为"森林之上的森林"，往往最高的树木 38～53 米高，如绿色的云悬挂在森林之上。上层是雨林之冠，高 20～30 米，树冠交织遮挡了下面植物的很多阳光，再下一层由较矮乔木的顶梢组成，高 12～18 米，再往下是更小的树和树苗，灌木在几乎得不到阳光的底层。因此用常规的丛林法去表现雨林的景色，也是力不从心的。

图 2-3 步步登高（博兰；雨林式造型）
Figure2-3 Ascending Step by Step (Poilaniella fragilis, Rainforest style)

A. It is not easy to convey the density and thickness of rainforest in artistic form.

B. It is hard to show the complex geographic features within a miniature landscape.

C. With a vertical structure, the tallest trees can reach to 38-52 meters. The upper part with interlacing branches is about 20-30 meters high, blotting out the sun; the middle part is composed of the crown of some lower trees, with a height of 12-18 meters; the lowest part is the brushwood with little sunshine. It is difficult to present such view in regular training way.

四、雨林式盆景的制作
Training of Rainforest Penjing

1. 雨林式盆景艺术形象的提炼 Creation of the artistic image

我们以大自然为范本师法造化，需要溶入作者的思想感情，融二神于一体，就是说"外师造化，中得心源"。不能机械地去模仿雨林，而要进行艺术上的构思、提炼、取舍，调动各种艺术、技术手段，去塑造我心中雨林的意象，使之成为"我的"雨林的艺术形象，传达我的思想感情（图 2-3）。

The source of creation includes both the nature and the designer's sentiment. We should not copy the rainforest mechanically, but should carry on artistic methods like reconstruction, refining, making appropriate choices and so on to create the image with thoughts.

2. 符合盆景艺术的创作规律 Creation Rules of penjing art

任何一门艺术都有它的规律和限定，"带着镣铐去舞蹈"是最具形象的说明。盆景艺术也有它的规律性，例如"缩龙成寸"，将一条巨龙浓缩在手掌之内，还要是龙的形象，绝对不能是虫、是蛇。小中见大、咫尺千寻，对于浓缩再现雨林景观，是一个极大的挑战。

Every form of art has its own rules and restrictions. So does penjing art. The minimized image must maintain the essence of its original shape.

3. 雨林式盆景的制作方法 Training method

（1）立意为先

雨林之中，景观纷呈，在盆景艺术创作规律的限制下，如何去表现雨林中的那些景致，取景是作者必须首先需要确定的，例如在绘画中都是画黄山，张大千和刘海粟就不同，梅清与弘仁也不一样。我表现雨林，突出的是阔叶树木生存竞争中的"争"，为此，调动一切造型手段去表现热带雨林中的"竞争的直线世界"（图 2-4）。

（2）选 材

在选材上我挑选萌发力强、易成活的杂木树种。以耐修剪、易蟠扎、成形快的树种为首选（如海南博兰等），常绿、观花观果等树种也是极好的素材，连根、连干、一本多干、卧干等桩胚在雨林式的造型中，往往起着事半功倍的效果。在雨林式盆景的选材上，尽量突出一个"野"字。

（3）定芽选枝

雨林式树木盆景造型时，依据未来的构图和造型，审视桩坯的芽点采取保留或舍去，靠接或催芽等措施，来达到有利造型的目的。雨林式造型选择树木枝条生长茂密的桩坯为好，利用天然的枝条造型，在造型上要注意：大与小、粗和细、长与短、疏和密等对立统一的美学法则，达到和谐统一的预期目的。

（4）枝干的造型

在雨林中，一颗古木倒地，在它的干身上又长出无数株树枝，最后都变成了无数株树木、树林，这是我创作雨林式盆景的自然现象中的依据。盆景艺术缩龙成寸，为此在造型上定位的枝，将来就生长成干、成树、成林。因为雨林树木向上直线生长是一个普遍的自然规律，所以在造型上力求枝干垂直向上。在造型的手法上，不拘一格，以达到目的为要求，绑扎和拉吊相结合，将枝条调直向上。

在雨林式盆景中，树木的造型还应该是以干为主，以枝为辅，这也是欣赏其他各式树木盆景的共同要求，在造型上还是要去掉一些过粗的枝条，调整枝与枝之间的穿插变化和修剪去掉多余枝和忌枝。

A. Determine a conception

Views in the rainforest emerge in great quantity. The first step for creation is to find a view to imitate. "A competitive world of straight lines" is what I try to present in my works.

B. Select materials

I prefer the trees with great vital power in the creation of Rainforest Penjing. The original trees should be easy for spurning, wiring and shaping. Besides, we may get twice the result with half the effort if trees with multiple, horizontal trunk, or connecting roots are chosen.

C. Choose buds and branches

Whether to save, remove or graft the sprouting points is guided by the designer's aiming image. Stumps with natural and thick branches are the best choice. Whereas the branches should have relative height, girth and density in order to compose harmonious landscape.

D. Train the branches

The natural view I try to imitate in miniature landscape is that branches grow from an old fallen tree, and finally become new trees. In the potted planting, branches are trained to grow straight upward to present the "new trees".

As well as other penjing styles, the shaping of Rainforest Penjing should give priority to the trunk, and then the branches. Consequently, over-thick and redundant branches must be removed.

Inspired by the mysterious natural forest, I do enjoy the process of producing Rainforest Penjing, and I will definitely keep pursuing perfection.

图 2-4 小雨林（博兰；雨林式初期造型）
Figure 2-4 Tiny Rainforest (Poilaniella fragilis)

第二节
雨林式盆景的作品展示
Works of Rainforest Penjing

- ◎ 作品名称：神奇的雨林
- ◎ 树种：博兰
- ◎ 石种：火山石
- ◎ 树高：180cm
- ◎ 盆长：200cm
- ◎ 心得：借助野性博兰桩材，将每根枝以垂直向上造型，以"枝"代树、欣欣向荣。原桩树干苍古雄奇、神秘莫测、盆面点石、缀其白石米以示水意，耐人寻味。缺点是枝条放长的粗度和高度还欠缺。

- ◎ Title of Work: Mystical Rainforest
- ◎ Tree species: Poilaniella fragilis
- ◎ Stone kind: Volcanic rock
- ◎ Tree height: 180cm
- ◎ Basin length: 200cm
- ◎ Creation experience: Each branch is trained upward as every single tree. The original tree is old and strong, lending mystery to the work. Furthermore, the decorative stones which stand for the water are quite thought-provoking.

◎ 作品名称：雨林世界（广东虫二居张华江先生收藏）
◎ 树种：博兰（一本多干）
◎ 树高：138cm
◎ 盆长：150cm
◎ 心得：依拙取势，以巧夺工。充分利用原始素材极不规范的树向，因势利导，创造一种较为奇特雨林世界。缺点是枝条走势尚需梳理。

◎ **Title of Work:** World of Rainforest (Collected by Mr. Zhang Huajiang, Guangdong)
◎ **Tree species:** Poilaniella fragilis
◎ **Tree height:** 138cm
◎ **Basin length:** 150cm
◎ **Creation experience:** Made the best use of the irregular branches and guided it to be a favorable factor, the work manifests a mysterious rainforest

◎ 作品名称：硫木亦成林
◎ 树种：博兰
◎ 石种：海母石
◎ 树高：72cm
◎ 盆长：150cm
◎ 心得：原桩是一件极普通的横生树木，通过谋篇布局，利用其提根悬根露爪；树干上留蓄众多芽点，养成枝干，形成一片森林，深化主题。盆面小溪点石，增其野趣。缺点是用盆需加宽，作品深度不够。

◎ Title of Work: A Wood with one Tree
◎ Tree species: Poilaniella fragilis
◎ Stone kind: Corallite
◎ Tree height: 72cm
◎ Basin length: 150cm
◎ Creation experience: The original stake used to be an ordinary diageotropic tree. While after training, part of the roots are exposed, and lots of branches have grown from the sprouting points, becoming a thick forest. Furthermore, the decorative stones make the work look quite natural.

◎ 作品名称：古林欢歌
◎ 树种：博兰
◎ 树高：188cm
◎ 盆长：180cm
◎ 心得：树木主体斜曲、昂首阔步，枝条形成多层次林片，主、次、高、低、大、小有别，参差错落、清新可人。缺点是用盆可再长一些，盆右边空间不够。

◎ Title of Work: Song of Ancient Forest
◎ Tree species: Poilaniella fragilis
◎ Tree height: 188cm
◎ Basin length: 180cm
◎ Creation experience: The major trunk of the tree is slantwise like marching forward, and the multilayer branches are in picturesque disorder.

- ◎ 作品名称：天涯新雨
- ◎ 树种：博兰
- ◎ 石种：海母石
- ◎ 树高：70cm
- ◎ 盆长：130cm
- ◎ 心得：这是一件较夸张的雨林式作品，是笔者有意将热带雨林树木移植海岛中间，并作近景处理，神形并茂、耐人寻思。

- ◎ Title of Work: Rain in Early Spring
- ◎ Tree species: Poilaniella fragilis
- ◎ Stone kind: Corallite
- ◎ Tree height: 70cm
- ◎ Basin length: 130cm
- ◎ Creation experience: It is a Rainforest Penjing with artistic exaggeration. The island with a forest in the middle dominates the foreground.

- ◎ 作品名称：南国雨林（海南奥林匹克公司收藏）
- ◎ 树种：博兰
- ◎ 石种：琼州石
- ◎ 树高：66cm
- ◎ 盆长：100cm
- ◎ 心得：这是一盆较典型树石结合雨林式作品，拿掉树就是山水盆景，去掉石便成丛林树木盆景。二者结合，则优势互补、巧夺天工。

- ◎ **Title of Work:** Rainforest in Southland(Collected by the Olympic Park, Hainan)
- ◎ **Tree species:** Poilaniella fragilis
- ◎ **Stone kind:** Qiongzhou stone
- ◎ **Tree height:** 66cm
- ◎ **Basin length:** 100cm
- ◎ **Creation experience:** It is a typical tree-rock Rainforest Penjing, combining the features of both landscape and forest penjing styles.

第三节
雨林式盆景的创作演示
Creation demonstration of Rainforest Penjing

一、雨林式盆景创作演示之一
Demonstration of Rainforest Penjing(1)

2006年1月9日,笔者在"海口市刘传刚盆景艺术中心"为海南大学观赏园林系的学生做了一场"雨林式盆景创作"示范表演。

On January 9th, 2006, Mr. Liu Chuangang gave a demonstration of Rainforest Penjing to the students of Landscape Architecture from Hainan University.

1

2

3

4

5

6

7

8

1. 表演用博兰树原貌；
2. 表演用石头为海母石；
3. 用盆为椭圆形大理石盆，长130cm；
4. 示范表演所用工具；
5. 将树木摘叶后每枝进行绑扎，并将每一枝条垂直向上造型（用金丝固定）；
6. 造型后的初步效果；
7. 对每一枝条再作精细调整；
8. 用专用工具对树木腐烂面进行加工；
9. 加工后将愈合剂涂在树木破损处，愈合并起防腐作用；
10. 将原树土球退土（去掉多余的原土，清理根系）；
11. 整理好根系后，重新放入大理石盆中开始布局；
12. 配置海母石（将海母石加工好后放入其中）；
13. 将每块海母石固定于盆中（用高标号水泥兑107胶水调和）；
14. 作品主景部分完成后的效果；
15. 作品完成后的全景图，该作品命名为——"天涯新雨"。

Figure 1. The original tree.
Figure 2. The selected stone: corallite.
Figure 3. The selected basin: an oval marmoreal basin with a length of 130cm.
Figure 4. Implements.
Figure 5. Pick the leaves, then wire the branches, make every branch to grow vertically (fixed by metal wires).
Figure 6. Half-finished work.
Figure 7. Further adjustment on each branch.
Figure 8. Deal with the decomposed parts with special implements.
Figure 9. Put some consolidant on the damaged parts to prevent decomposing.
Figure 10. Clear away the redundant soil on the roots.
Figure 11. Replant the tree in a new basin.
Figure 12. Add some processed corallite.
Figure 13. Fix each stone in the basin with high quality cement and 107 glue.
Figure 14. General look of the work.
Figure 15. Finished work named "Rain in Early Spring".

二、雨林式盆景创作演示之二
Demonstration of Rainforest Penjing(2)

2012年3月30日，应海南花卉大世界"盆景之友会"之邀，笔者在锦园"盆景艺术教研室"做了一场"雨林式盆景创作示范表演"。

Invited by the Hainan Flower World, Mr. Liu Chuangang put on a demonstration of Rainforest Penjing on March 30th, 2012 in his garden.

1. 表演树木——博兰，枝条高度为216cm；
2. 表演所用工具；
3. 表演用盆为长方形紫砂盆，长110cm，宽65cm；
4. 树木摘叶后，修剪枝条，将枝条高度控制在120cm以内；
5. 清掉多余的枝条；
6. 对每根枝条进行绑扎；
7. 枝条绑扎后的局部效果；
8. 脱盆。将绑扎和造型后的树木从原盆中脱起，并修剪根系；
9. 将铁纱网放入盆洞中；
10. 将造型好树木放入盆中（放置树木前，盆中要放置好盆土）；
11. 细心将盆土压紧；
12. 开始铺种苔藓；
13. 作品完成后的效果。

Figure1. The original tree: Poilaniella fragilis, the length of its branches is 216cm;
Figure2. Implements;
Figure3. The selected basin: an oblong boccaro basin with a length of 110cm and a width of 65cm;
Figure4. Pick the leaves, then prune the branches within 120cm;
Figure5. Clear away the redundant branches;
Figiure6. Wire every branch;
Figure7. The wired branches;
Figure8. Get the tree out of the original basin and prune the roots;
Figure9. Lay a gauze screen on the hole of the basin;
Figure10. Put soil in the basin and then replant the tree;
Figure11. Press the soil;
Figure12. Put on the moss;
Figure13. Finished work.

1

2

3

4

5

6

7

8

9

10

11

12

13

三、雨林式盆景制作要领

如痴如醉入雨林，魂牵梦萦不了情。
依题选材重奇字，野乱杂怪方称心。
轰然倒塌千古树，龙蛇起舞闹纷纷。
一枝亦成一棵树，一树又为一片林。
密密匝匝遮天日，直起直落争生存。
疏密聚散偏于密，顾盼争让重在争。
盆中抒情胸中志，点点滴滴为创新。

第三章
树木盆景创作探索
A Survey of Tree Penjing

第一节
树木盆景造型的主要形式
Main Styling Forms of Tree Penjing

在盆景艺术造型中用形式来分类，是一个求简求同的过程，以把握树木类型的共性特征，便于在千姿百态的树木中去提炼精要，来进行对树木的造型。在造型过程中，不能仅仅关注树木形式上的共性特征，更重要的是认真地挖掘各个桩坯与众不同的特点，在形式中去创造、塑造个性鲜明的树木形象，使本来已经形式化类型化的树木形象独特，塑造的树木形象寓意深刻而又丰满，这是一种追求个性、求异、寓繁于简的创作过程（图3-1）。

To classify the shapes of miniature trees by form is a process of simplification and stylization, summing up the general features and selecting the essence of various trees. During the training, it is important to rediscover the specificities of different original trees, and to create distinctive images that pregnant with meaning on the basis of a certain form(Figure3-1).

一、直立式树木盆景的造型
Training of upright penjing style

直立式树木好像人站着，总的原则是在静中求动、稳中求险、平中求奇。直立式的树木造型需要突出地做好一枝主枝，如大飘枝、跌枝等，以枝的动感，打破或缓和主干的呆板、僵硬、无变化等不利因素，并求得在对立统一的美学规律中矛盾双方的互补，在造型上使作品展现为一个生动的姿态。直立式树木盆景也有多种结构形式，如独干、双干、

图3-1 翠云飞舞（香兰）
Dancing Green Clouds(Chinese cymbidium)

多干，还包括曲干、丛林等（图3-2至图3-4）。

直立式树木盆景要选择那些根部四面展开、八方辐射的桩坯，在造型上才能"站得住"，有"牢固感"和"安全感"。当然，选择是一个方面，用后天的根部造型以弥补先天的不足，则是努力的方向。还有，就是盆面上的主根或较粗的根，要与飘枝、跌枝的走势大体一致。

直干式树木的造型口诀为：

直干高耸，分枝平垂，力求俯枝，平中求奇。

The leading elements of upright penjing style are being static, steady, orderly, blending with some flexible motion. Prominence should be given to a principal branch in the training of upright penjing style, such as a big waving branch, a declining branch and so on, whose vividness can alleviate the flat and stiff feeling caused by the trunk. Upright penjing style can be subdivided into different kinds, like single trunk, double trunk, multi-trunk, slanted trunk, forest style and so on. (Figures 3-2 to 3-4)

Original trees of roots growing on all sides are the optimal choices to produce upright style penjing, since they can provide a stable look. Further more, acquired training could make up congenital deficiency. On the other hand, the principal or thicker surface roots in the pot should be largely consistent with the moving direction of the waving and declining branches.

The rules of training upright penjing style are summed up as following:

Upright trunk stands tall and erect; side branches smooth out horizontally; bending branches add flexibility.

二、倾斜式树木盆景的造型
Training of slanting penjing style

如果形容直立式树木如人站立，那么倾斜式的树木就如人或走或跑、或倚或舞、或跳或跃的感觉。倾斜式树木盆景应该是在树木盆景中为数最多、最为普遍的一种形式，是变化最为丰富的一种形式，所谓"一斜则千变万化"（图3-5至图3-10）。

分析树木倾斜的形成原因：一是外因，即树木所在环境；一是树木的内因，与树木的趋光性、向上性有着密切的关联，例如树木因雪压霜欺、雨暴风狂、山体滑坡、土壤松动等，外力的作用造成倾斜，树木继而顽强的生长，由倾斜而产生弯曲；树木或生长于斜坡陡壁，或面临池塘湖泊，因而争向空旷、阳光水气，甚至侧枝长成主干等诸多原因，形成树木的主干倾斜。

倾斜的本身就意味着运动。在直立式树木造型中要静中求动，稳中求险，平中求奇；倾斜式树木的造型则正好相反，要分析那些不稳定的因素，要险中求稳，动静相宜，奇正相依，以达到均衡的艺术效果。

当树木受到外力而造成倾斜时，有些树根会被折断或者弯曲，天长日久，这些弯曲的树根粗壮后，就能形成一种极具力度的根爪。将这种根爪的造型运用在倾斜式树木的造型中去，抓撑着倾斜的树干，能起到调节视

图3-2 直立式单干树木盆景
Figure 3-1 Single-trunk upright penjing

图3-3 直立式双干树木盆景
Figure 3-2 Double-trunk upright penjing

图3-4 直立式丛林盆景 Forest-style upright penjing

图3-5 微斜式树木盆景 Slanting style

图3-6 倾斜式树木盆景 Slanting style

觉感观以达到均衡的效果。

倾斜式树木的主干由于倾斜的程度不同，所引起视觉上的动感、力度和气势也是不同的，又由于不同形态的主干倾斜的程度上所引起的种种不同变化，则是造成"一斜则千变万化"的原因吧！

倾斜式树木的结顶，能起到调节树木因倾斜所造成的不稳定因素，故结顶左一点或右一点都可能影响树木的动势或稳定。因此，对倾斜式树木造型的结构定位，一定要认真地斟酌和推敲，准确地确定结顶的位置，对树木造型完美地表达造型目的，有着重要意义。

倾斜式树木枝托的分布，要依据立意构思和桩坯的实际情况而定。但有一点应该把握，那就是顺势的枝托，能加强倾斜方向的动势，逆势的枝托可以起到均衡，以达到因倾斜造成不安的感觉。

在倾斜式树木造型中，还有一种回头的造型，其主干倾向一边，然后"回头"弯曲转折，结顶和主枝倾斜向相反的方向，这种造型非常生动，于老辣中见活泼，雄浑中见潇洒，秀丽中见灵动，是一种令人喜爱的造型形式。

倾斜式树木的造型口诀为：

斜曲多姿、几见波折、动势飞扬、力求均衡。

Upright penjing style is compared to a standing man, while slanting style is like a man in motion. It is the most common style with an appreciable amount and great variety. (Figure 3-5 to 3-10)

There are several reasons for slant: one is internal cause, like growing upward and toward the sun by instinct; the other is external cause, the tree's growing environment, such as bowed by snow, blew by wind, landslide, loose soil and so on. Trees compete and struggle to survive under the changing conditions, and sometimes the side branches may even grow into a principal trunk and become slanting.

There is a popular form of slanting penjing style, whose top branch lean to the opposite side of the trunk, shaped like turning around. This form is vigorous and natural, elegant and agile.

The rules of training slanting penjing style are summed up asfollowing:

Winding trunk of various stances; branches with twists and turns; dynamic feeling occupies the leading position while balance should not be neglected.

三、悬崖式树木盆景造型
Training of hanging cliff penjing style

悬崖式树木盆景表现悬崖峭壁上横倚或倒挂着生长的树相。由于树木生长在绝壁之上，凌空倒悬，与云海为伴，与风涛博击，傲视苍穹，力拍云天，可以表现那种不屈不挠，顽强拼搏的精神。

横倚式树木在造型上是介于倾斜式与悬崖式两者之间的过渡形式，若树木的主干向上延伸就进入倾斜式，如主干向下生长则过渡到悬崖式树木的造型范围。一般来说树木的主干卧于盆内称为偃卧式，横于盆外称为横倚式，俗称小悬崖，如果主干跌宕向下，树梢向盆底或低于盆底，俗称大悬崖（图3-11至图3-14）。

悬崖式树木造型选择桩坯时，三曲有致，起伏跌宕，

图3-7 倾斜式双干文人型 Double-trunk slanting style

图3-8 倾斜式临水型 Facing-water slanting style

图3-9 倾斜式回头造型 Turning-around slanting style

图3-10 倾斜式树木主枝悬出盆外的变化造型 Developed slanting style

顿挫分明、气势流畅的桩坯当为首选，根理部分不仅需要遒劲有力，而且需要有一个曲度，否则树木就不好"挂"下去。有些悬崖的树木盆景将根理都搁贴在盆口上，显得软绵绵而缺乏力度和精神。因此悬崖式树木的造型对桩坯根理的要求需要有"咬定青山不放松"和"根深何惧临悬崖"的意味和神韵。

悬崖式树木的枝托造型因枝梢向下，所以失去了顶端优势，因此要控制上部分枝托的长势，放长下部分的枝托，以达到粗壮的目的。还有一个办法是：在养护生长期间，将盆横卧放置，变底枝成为顶端优势，待其枝托达到一定粗度后，树木自身的营养分配形成以后，再将盆立起来，枝托的长势依然会很好。

悬崖式树木的造型或横空出世、凌空飞渡，或如虬龙傲游、金蛇狂舞，险峻而又磅礴，因此造型上应注意把握险中求稳、动静相依的原则，去表现蟠屈回旋、飘逸多姿、顿挫分明的气势。由于主干向下生长、而枝条生长的向上性，在这"上""下"的矛盾中，要在整体上去取势，"上""下"的矛盾绝对不能太突出，也就是说，向下生长的主干上不要有太向上长得很高的主枝，否则就破坏了作品的气势，也显得极不自然。

悬崖式树木造型的口诀是：

悬崖险峻、挂而不卧、三曲有法、险稳结合。

Hanging cliff penjing style shows the look of a tree living on a steep cliff and growing horizontally or downward. Hanging upside down from the precipice, the trees may keep clouds company in sky or battle against winds and waves, representing the ambition of being prosperous.

Horizontal style is situated between slanting style and hanging cliff style. If the horizontal trunk grows upward, it will become slanting style, and downward become hanging off style. Generally speaking, the tree with its trunk lying down inside the pot is lying trunk style, outside the pot is partial hanging cliff style, while with a rapidly descending trunk growing below the lip of its pot is called hanging cliff style. (Figures 3-11 to 3-14)

The original tree with undulating twists and turns in sharp contrast is the first preference for creating hanging cliff penjing. The tree should grow upright with a strong root-flare for a small distance but then bend downward. Otherwise it can be difficult to maintain a downward-growing tree. Without this distance, it will lose the implication of a tenacious struggle against adverse circumstances.

The rules of training hanging cliff penjing style are summed up as following:

Hanging downward from the precipice; fluctuating with twists and turns; combining adversity and steadiness.

四、丛林式树木盆景的造型
Training of forest penjing style

丛林式树木盆景是表现大自然中树林的一种形式。在我们的星球上有各种各样的树林：如莽莽苍苍的原始森林、神奇莫测的热带雨林、大漠中悲壮苍凉的胡杨林、海滨伴随着蓝天碧浪的椰树林……不同的地理环境，赋予不同的树林不同的典型特征，正是这些典型特征，是盆景创作者必须掌握的创作丛林式盆景的钥匙。

图3-14 倚式小悬崖（海底捞月式）Partial hanging cliff style

图3-11 倚式小悬崖树木盆景 Partial hanging cliff style

图3-12 三曲有致悬崖树木盆景 Hanging cliff penjing with undulating twists and turns

图3-13 悬崖双干树木盆景 Hanging cliff penjing with double trunks

丛林树木盆景的用材，以相同的树木组合最为常见，它的优势是作品的整体感容易把握和体现，养护上也相对方便些。要选择格调和谐一致的桩坯，并且还要注意这些树木的发芽时间、叶片大小、发芽的颜色一致等。用不同树种结构的丛林盆景，可以表现十分丰富的内容，例如中国盆景艺术大师贺淦荪先生的丛林盆景《秋思》就是用了榆、三角枫、朴、牡荆、水腊五个不同的树种，结构成一幅立体的画，统一于"西风"之中（图3-15）。不同树种的组合丛林，首先要注意把握主次关系，确立主树、副树、客树等之间的关系，既不要喧宾夺主，又不能平均罗列。多树种组合还要充分考虑不同树种的生理和生长规律，依据不同树种对光照、水肥、抗旱耐湿包括对酷暑严寒的忍耐程度作科学合理的统一安排，让不同的树种植物共处一盆之中，能够正常的生长。

"三远法"在丛林盆景中是能够体现的，仰视树林主树显得高大，俯视树林树与树的高矮变化甚微，由前往深处观察树林层次丰富（图3-16至图3-18）。

一盆丛林盆景，无论有多少株树木，总是以三株树木的组合为开始，称为三树组织法。多株树木的组织和变化，寓于三树组织法之中。在三树组织中，首先要确定主树、副树和客树，主树高大粗壮，依次副树和客树，否则就主次不分了。一般来说三株树木之间的粗细高矮不宜过分悬殊，不然三树的构成就不像树林之间的关系，而成为树木远近之间的感觉。

三树组织时，不要将主树栽植在盆的正面和侧面的正中位置，也不要栽植在盆的边缘的位置，最恰当的位置在盆正面或左或右的三分之一处，盆的侧面二分之一的偏前或偏后的位置上。要注意主树的走势，一般来说，走势向左置入盆的右方，向右置入盆的左方。主树定位后副树和客树的位置依据它们的形态见机取势而定，以

三株树木的形状、神态在整体上达到相随相伴、相从相盼结构成一个整体，而不是相背相离、各自为阵、相弃相离、貌合神离的局面。

三株树木的栽植点以结构成不等边三角形为宜，使二株树聚，一株树散，不要将三树形成等边三角形和直角三角形，因为只有不等边三角形极具动势和形成活泼自然的格局，能使轻重、疏密、聚散、争让、顾盼等一系列对立统一美学法则得以体现，无论是俯视、平视或仰视，尽力让树木构成不等边的三角形，结构成错落有致、和谐大方的格局。

多树的组织以三树组织为基础，如五树的组织可在主树和副树的附近各增添一树，七树的组织以可以在五树的基础上在主树与客树的附近再各增添一树，如此类推。多树组织的要领是：①注意主树在整体上的分量，不可丧失主树的气势和重量感；②无论有多少树或分成几组，都要把握住一个视觉中心，在争让中见到顾盼的关系，而不能各自为阵"另立山头闹分裂"；③把握住起"秤砣"作用的客树位置的分量，注意在视觉上达到协调美；④尽量能做到让盆中的任何三株树木的关系，都能构成不等边的三角形，任何两树的关系，不与盆边形成平行线；⑤把握虚实关系，不仅树木植物需要通风采光，在视觉上也需要布白留空，处理好聚散和疏密等关系。

在一组丛林树木的组织中，以三、五、七、九等奇数为好，比较容易分布和搭配。如果丛林盆景是由两组或三组丛林来组成结构，那么树木的数量是奇数或偶数就不重要了。例如由五株树木为一组，和三株一组的树木来搭配成一盆丛林就十分合理。另外有些树木是双干的，或者是多干，总之具体情况具体分析，以达到合理的组织，而不要拘泥概念上的定义（图3-19至图3-21）。

图3-15 多树种组合丛林盆景（《秋思》贺淦荪作）
"Autumn Thoughts" contains different species of plant material, He Gansun

图3-16 丛林盆景高远法取景，仰视树林主树显得高大 Viewed from the bottom, the major tree is stronger than the rest

图3-17 丛林盆景平远法取景，树与树高矮变化不大 Viewed from the top, there is almost no difference in height

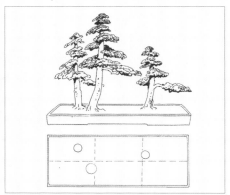

图 3-19 丛林盆景三树组织法，主树的位置在盆三分之一处，侧视二分之一偏前或偏后，三棵树组合为不等边三角形。The method of three-tree combination. The planting points of the three trees should present an outline in the form of a scalene triangle.

图 3-20 在三树组织的基础上增添两棵树便组成五树丛林。The arrangement of five trees is on the basis of three-tree combination, adding two trees by the side of the three.

图 3-21 在五树组织的基础上再添植两棵树就组成了七树的丛林，不论是五树丛林还是七树丛林，将其中任何三棵树连结都能构成不等边三角形。Adding two trees on the basis of five-tree combination, it becomes a seven-tree forest. The structure of an equilateral triangle should be avoided in every three-tree composition.

丛林盆景造型的口诀是：

丛林多景、虚实相宜、争让顾盼、协调统一。

The compositions of forest style penjing usually contain a single species, which give full play to integral perception and are relatively easy to maintain. In this case, the original trees should be of a single rhythm and coincident in the time of sprouting, sizes of blades, colors of buds and so on. On the other hand, rich content could be displayed if multiple species are used. For instance, the work named "Autumn Thoughts" (Figure3-15) contains five different species of plant material. sTypically, in the compositions of multiple species, a hierarchy is developed with one dominant tree and each other subordinate in turn. Scientific and reasonable arrangement should be taken according to the growing habits of different species in the same container, such as their requirement for sunshine and manure, and their resistance of humidity and temperature.

Application of "Three Ways of Perspective" can be found in forest penjing style. When viewed from the bottom, the major tree is stronger than the rest; viewed from the top, there is almost no difference in height; viewed from the front, distinct gradation can be seen. (Figures 3-16to3-18)

No matter how many trees there are in a container of forest style penjing, arrangement should begin with three of them. That is called method of three-tree combination, which provides the basis of combination of multiple trees. The priority for arranging trees in a group planting is to consider their height and girth. The major tree should be highest and strongest, next the secondary and finally the minor. Nevertheless, great disparity needs to be avoided since the planting is designed to convey a natural image of a forest group.

The planting points of the three trees should present an outline in the form of a scalene triangle instead of an equilateral or a rightangled one, with two trees stand closer, while the other with a relative distance. By this kind of structure the planting can give expression to a dynamic, lifelike and natural style. As a result, the designer can arrange the trees to form a series of scalene triangles viewed from different aspects.

The arrangement of multiple trees is on the basis of three-tree combination, adding trees by the side of the three.

The number of trees in a single group should be an odd, whereas it is not a hard and fast rule, especially when trees of doubletrunk or multiple-trunk are used. (Figures 3-19 to 3-21)

The rules of training forest penjing style are summed up as following:

To create complex scenery with distinct perspective, conveying a natural and harmonious relationship.

图 3-18 丛林盆景深远法取景，树林层次丰富
Viewed from the front, distinct gradation can be seen

第二节
树木盆景的作品展示
Works of Tree Penjing

- ◎ 作品名称 阅尽人间春色（全国金奖、香港郑在权先生收藏）
- ◎ 树种：博兰
- ◎ 树高：118cm
- ◎ 盆长：98cm
- ◎ 心得：该作品是笔者树木盆景代表作之一。作品选材极具挑战性，利用树木双干古朴奇特之貌，见机取势，巧夺天工；树木造型因势利导、平奇互用、动势飞扬。缺点是右边飘枝粗度尚需加大，力度还欠缺。

- ◎ Title of Work: Passing through the Spring
- ◎ Tree species: Poilaniella fragilis
- ◎ Tree height: 98cm
- ◎ Basin length: 70cm
- ◎ Creation experience: This is one of Mr. Liu's representative works. It is such a challenging task to train this old double-trunk tree, which shows art beats nature. However, the flowing branch on the right should be stronger.

- ◎ 作品名称：将军风采（全国金奖作品）
- ◎ 树种：博兰
- ◎ 树高：108cm
- ◎ 盆长：80cm
- ◎ 心得：这是一盆较规矩的树木盆景大树型造型，根盘向四周伸展爬地而起，根理健壮。收尖渐变，给人以稳重感；枝法造型：注重穿插、变化，参差错落、呈不等边三角形。缺点是用盆稍偏大，根盘透空处放石头可能效果更佳。

- ◎ **Title of Work:** Images of General
- ◎ **Tree species:** Poilaniella fragilis
- ◎ **Tree height:** 108cm
- ◎ **Basin length:** 80cm
- ◎ **Creation experience:** It is a work of tree-shape penjing with strong roots and gradually-changing top. The branches are in picturesque disorder and form a shape of equilateral triangle. Whereas, the chosen basin is a little bit big, and stones should be added at the roots to make it perfect.

- 作品名称：贵妃出浴
- 树种：朴树
- 树高：96cm
- 盆长：108cm
- 心得：利用树木天然形态进行自然布局。同时巧借树木身段拟人化，增强艺术感染力,。缺点是枝干造型尚缺年功。

- **Title of Work:** Immersed in Spring Breeze
- **Tree species:** Hackberry
- **Tree height:** 96cm
- **Basin length:** 108cm
- **Creation experience:** This work makes good use of the tree's original shape, but the branches still need further training.

- ◎ 作品名称：迎宾图（国际园博会银奖作品）
- ◎ 树种：香兰
- ◎ 树高：120cm
- ◎ 盆长：85cm
- ◎ 心得：这是一株仿松造型的香兰树。树干直立而挺拔苍劲，云片布局舒展有度，层次活泼且富于多层次变化，是香兰盆景造型中较典型的树象之一。缺点是左边大飘枝太直了点，若有些弯曲再转直线会更好。

- ◎ **Title of Work:** Greeting Guests
- ◎ **Tree species:** Chinese cymbidium
- ◎ **Tree height:** 120cm
- ◎ **Basin length:** 85cm
- ◎ **Creation experience:** The Chinese cymbidium is trained into the shape of a pine tree with a tall and straight trunk. The leaf-layers are unfolding in good proportion. Whereas, the flowing branch on the left is lack of curve.

◎ 作品名称：六顺歌
◎ 树种：博兰
◎ 树高：98cm
◎ 盆长：70cm
◎ 心得：这是一株一本六干连体博兰树，该材料比较难得，树干主次高低，错落有致。缺点是配盆偏大，根盘尚需梳理。右边第一干中需有一个较粗壮的飘枝，以达构图效果。

◎ Title of Work: Good Luck with Six
◎ Tree species: Poilaniella fragilis
◎ Tree height: 98cm
◎ Basin length: 70cm
◎ Creation experience: The six-trunk tree is quite rare for the training of penjing. This work makes a distinction between the important and the lesser, and the branches and leaves are well-spaced. Nevertheless, the selected basin is too big, and the roots still need to prune. Furthermore, it will look better if there is a strong flowing branch on the first trunk on the right.

◎ 作品名称：古林春色（广西梁宏文先生收藏）
◎ 树种：博兰
◎ 树高：88cm
◎ 盆长：180cm
◎ 心得：这是一盆大型组合树木盆景。作品选用多株博兰树进行组合。树木主、次、高、低、大、小有别，但均以直线为基调，盆面点缀海母石增其构图效果与盆面变化。缺点是枝条密度不够。

◎ **Title of Work:** Spring of Ancient Forest (Collected by Mr. Liang Hongwen, Guangx)
◎ **Tree species:** Poilaniella fragilis
◎ **Tree height:** 88cm
◎ **Basin length:** 180cm
◎ **Creation experience:** It is large-size work of tree pengjing. Trees in various sizes are used to compose a work that featured at straight lines. On the other hand, ornamental stones are added to make it look more natural.

◎ 作品名称：鹿回头（全国银奖作品）
◎ 树种：博兰
◎ 树高：68cm
◎ 盆长：70cm
◎ 心得：这是一棵较难得的象形博兰树木，似奔鹿至天涯而回头，见证了三亚"鹿回头"的爱情传说故事。作品富于情感、章法独具，小中见大，动静结合。缺点是用盆略偏小，偏浅，枝法层次还需再疏朗些。

◎ Title of Work: Deer's Turn
◎ Tree species: Poilaniella fragilis
◎ Tree height: 68cm
◎ Basin length: 70cm
◎ Creation experience: Named from a love story in Hainan Island, the shape of this work just like a deer's turn-about. It has an unusual composition while the basin is a little bit small and shallow, and the layers of branches should be clearer.

- ◎ 作品名称：蹉跎岁月
- ◎ 树种：博兰
- ◎ 树高：120cm
- ◎ 盆长：100cm
- ◎ 心得：这盆博兰树上盆造型20年，树干苍古雄奇，历尽桑田。枝条造型虬曲多姿，可见年功。缺点是大飘枝所留位置欠佳，粗度也不够。

- ◎ Title of Work: Idling away
- ◎ Tree species: Poilaniella fragilis
- ◎ Tree height: 120cm
- ◎ Basin length: 100cm
- ◎ Creation experience: It is a work which has been trained for 20 years. The trunk looks impressively old, and the branches are rich in twists and turns. Nevertheless, the largest flowing branch should be stronger and in a better position.

◎ 作品名称：小鸟天堂（海南冯川建先生收藏）
◎ 树种：博兰
◎ 树高：88cm
◎ 盆长：65cm
◎ 心得：这株博兰树上盆造型19年，经过多年剪枝造型，可谓枝繁茂盛。该树木为一本多干，参差错落有序，深化主题思想。缺点是配盆偏大、偏深，右边飘枝力度欠缺，中间主树不够明显，若再能加粗加高效果一定不同。

◎ **Title of Work:** Bird's Paradise (Collected by Mr. Feng Chuanjian, Hainan)
◎ **Tree species:** Poilaniella fragilis
◎ **Tree height:** 88cm
◎ **Basin length:** 65cm
◎ **Creation experience:** It is a branchy work which has been trained for 19 years. It uses a multi-trunk tree to create a well-spaced shape. However, the selected basin is too big and deep. And the flowing branch on the right should be stronger. Further more, it will look much better if the trunks in the middle are taller and stronger.

- ◎ 作品名称：平步青云
- ◎ 树种：香兰
- ◎ 树高：188cm
- ◎ 盆长：110cm
- ◎ 心得：作品可谓古木新姿，树虽老而枝繁叶茂。云片造型讲求多层次变化而且步步登高，紧扣主题。缺点是树木层次不太清晰。

- ◎ Title of Work: Meteoric Rise
- ◎ Tree species: Chinese cymbidium
- ◎ Tree height: 118cm
- ◎ Basin length: 110cm
- ◎ Creation experience: It is an old tree with luxuriant foliage. The leaf-layers gradually rise to the top. On the other hand, the small flowing branch on the lower right should be removed.

◎ 作品名称：独秀
◎ 树种：香兰
◎ 树高：110cm
◎ 盆长：55cm
◎ 心得：该树木造型简洁明快、潇洒自如，特别是顶部一飘枝打破常规，破平立异，形成艺术个性。缺点是树木下部右边云片偏大，有些小枝尚需梳理。

◎ Title of Work: Outshine Others
◎ Tree species: Chinese cymbidium
◎ Tree height: 110cm
◎ Basin length: 55cm
◎ Creation experience: It is a neat and sprightly work featured at the flowing branch on the top, which breaks away from conventions. However, the leaf-layers on the lower right are a little bit large, and some of the branches still need to prune.

- 作品名称：绿林春野（广西梁宏文先生收藏）
- 树种：博兰
- 树高：86cm
- 盆长：180cm
- 心得：这是一件大型博兰组合丛林式树木盆景。主树靠右形成主导地位，配树和远景树大小适宜，顾盼有情，盆面处理自然起伏，浑然一体，烘托和深化了主题。缺点是若用盆再宽一点，景的深度还可以加大，意境将更深远。另外树木枝条还欠丰富。

- **Title of Work:** Green Forest in Spring
- **Tree species:** Poilaniella fragilis
- **Tree height:** 86cm
- **Basin length:** 180cm
- **Creation experience:** It is a large-size work of combination tree penjing. Trees on the right dominate the others. The ornamental stones help to create ups and downs, making the work look more natural. Whereas, it will be better if the basin is larger and deeper.

第三节
树木盆景的创作演示
Creation demonstration of Tree Penjing

一、树木盆景创作演示之一
Demonstration of Tree Penjing(1)

2001年9月11日，应中国盆景艺术家协会之邀，笔者在成都"全国第十三届盆景与园艺培训班"上做了一场树木盆景教学示范表演。

Mr. Liu Chuangang showed how to make a tree penjing in Rizhao for the 13th training class of penjing & horticulture.

1 2 3
4 5

1. 表演树木为成都武侯祠提供的六月雪，该树木从未造型过，笔者选择两株六月雪做成垂枝式树木盆景；
2、3、4. 将每根枝条细心绑扎，改变原向上生长习性而变成垂枝形态。
5. 边绑扎边修剪造型，去掉多余的枝条和芽点；
6. 造型后上盆（点石和铺植好苔藓）；
7. 作品完成后，笔者向观摩者介绍整个创作过程，将作品命名为"春风杨柳"；
8. 笔者和观摩者在作品旁合影留念。

Figure1. The selected trees;
Figure2,3,4. Wire the branches to grow downward;
Figure5. Prune the trees and remove redundant branches and sprouting points;
Figure6. Replant the trees, then add stones and moss;
Figure7. Explain the training process. Finished work: Willow in the Spring Breeze;
Figure8. Group photo.

6

7

8

二、树木盆景创作演示之二
Demonstration of Tree Penjing(2)

2003 年 10 月 2 日，应中国盆景艺术家协会之邀，笔者在山东日照举行的"全国第十四届盆景与园艺培训班"上做了一场树木盆景教学示范表演。

Invited by the Chinese Penjing Artists Association, Mr. Liu Chuangang gave a demonstration of tree penjing in Rizhao for the 14th training class of penjing & horticulture.

1. 表演树木为当地所提供的小叶榆，基本属未造型小树木。示范表演前摘叶（培训班学员协助表演）；
2、3. 对每株树木进行修剪整形（修枝、修根等）；
4. 笔者对整形后的树木向学员作讲解；
5. 上盆前将铁纱网固定于盆底中；
6. 培土。将盆景土填放均匀；
7. 安植树木，先固定好主树（指组合树木盆景中为主的树木）；
8. 再配植次树，使其相互协调；
9. 细心将所配树木用土壤压实和稳固；
10. 点缀山石（龟纹石），处理好盆面，使盆面高低起伏、凹凸不平，符合自然的效果；
11. 作品完成后，笔者向学员讲解整个创作过程，并现场为作品命名为"群英进发"。

Figure1. Pick the leaves of the selected trees with the help of some trainees;
Figure2,3. Prune the branches and roots;
Figure4. Explain the process;
Figure5. Lay a gauze screen on the hole of the selected basin;
Figure6. Put soil in the basin and make it even;
Figure7. Replant the major tree in the basin;
Figure8. Replant the subordinate trees;
Figure9. Earth up carefully;
Figure10. Put more soil and stones to create ups and downs;
Figure11. Finished work: Setting out.

三、树木盆景创作演示之三
Demonstration of Tree Penjing(3)

2005年9月8日，应"第八届亚太盆景赏石会议暨展览会"组委会之邀，笔者在北京植物园做了一场树木盆景专场示范表演。

Invited by the Committee of the 8th Asia-Pacific Penjing & Shangshi Exhibition & Convention, Mr. Liu Chuangang put on a demonstration of tree penjing in Beijing arboretum on September 8th, 2005.

1. 所表演树木为海南博兰树（右为笔者夫人王四英为助手）；
2. 树木摘叶后的效果；
3. 修剪树木多余的枝条进行修剪；
4. 将树木每一根枝条绑扎（用大小不同的金属丝顺时针绕开芽点进行绑扎）；
5. 绑扎完后的全景效果（将原树木生长的形态进行了彻底改变，原树木枝条全部向上自然生长，现改为向下垂直生长，此形式为垂枝式盆景造型）；
6. 笔者向观摩者讲解树木造型创作过程；
7. 所选择盆为长方形紫砂盆，尺寸为70cm×30cm×19cm（长 × 宽 × 高）；
8. 将树木放入盆中定植（中为云南省盆协盆景爱好者协助创作）；
9、10. 表演完后的作品全景效果。笔者再次向观摩者介绍创作的全过程，并将作品命名为"情满京城"，表达作者及海南盆景界对北京举办此次国际盆景活动的深情厚谊。
11. 表演结束后，北京植物园党委书记张佐双（右 2）、湖北李斌（右 3）、安徽张志刚（右 4）和笔者在作品旁合影。笔者将该作品赠送给北京植物园作永久收藏。

Figure1. The selected tree: Poilaniella fragilis (Assistant: Mrs. Wang Siying);
Figure2. The tree without leaves;
Figure3. Prune the branches;
Figure4. Wire the branches in clockwise direction bypassing the sprouting points;
Figure5. The wired tree. (The branches are wired to grow downward);
Figure6. Explain the training process;
Figure7. The selected basin: an oblong boccaro basin, 70×30×19cm;
Figure8. Replant the tree in the selected basin with the help of a fan from Yunnan Bonsain Association);
Figure9,10. Finished work: Affection for Beijing;
Figure11. Group photo: Liu Chuangang, Zhuang Zuoshuang, Li Bin, Zhang Zhigang (From right).

9

10

11

四、树木盆景制作要领

树分四歧有长短，势取向背弛亦张。
协调轻重向势枝，强化动势枝蓄长。
欲伸先缩背势枝，先退后进枝转向。
前枝遮掩层次丰，后枝陪衬看四方。
枝见波折分短长，三枝聚散须露藏。
渐变收尖看变化，主干苍劲根理壮。
不拘一格求神韵，以形写意势飞扬。

第四章
山石盆景创作探索
A Survey of Landscape Penjing

第一节
山石盆景的主要形式与分析
Main Styling Forms and Analysis of Landscape Penjing

山石盆景是极具中国盆景艺术特色的重要盆景种类，承载着中国艺术传统的美学思想，也表现了今天人们与大自然和谐相处的渴望与追求。中国地大物博，能制作山石盆景的石种俯拾皆是。忘不了20世纪80年代，我在鄂南的山山水水中，调查和寻觅，发现能制作山石盆景的石种有二十多种。不同的石种，具有不同的石质、纹理和色彩，也就是说它们具有创作山石盆景的不同艺术语言。忘不了当时的创作激情，几乎每天都要做上好几盆山石盆景。

山石盆景整体的造型要因材制宜，在一组山石的材料中，要熟悉每块材料的形态和安排它们的用途，制作时要做到"石质相同、石色相近、石纹相似、脉理相通"。在石料的加工上要做到顺乎自然、巧夺天工，虽由人作、宛若天成。

Landscape Penjing, an important style that enriched with artistic features of Chinese penjing, incarnates the aesthetic ideology of Chinese artistic tradition, and manifests the aspiration for harmonious relationship between human and nature. China is a big country abounding in natural resources, including the raw materials for Landscape Penjing. In the time of 1980s, I found more than 20 kinds of potential rocks various in texture, grain and color. With

图 4-1 峰状山石盆景造型 Peak-shape Landscape Penjing

图 4-2 岩状山石盆景造型 Cliff-shape Landscape Penjing

great enthusiasm for creation, I made works of Landscape Penjing almost every day.

The whole styling of Landscape Penjing must be in line with specific materials. We should be familiar with each rock to make appropriate arrangement following the principle of being same in texture, similar in color, and linked in grain. Although sometimes the raw materials need to be processed, they should be natural in appearance.

一、峰状山石盆景造型
Peak-shape Landscape Penjing

峰状山石盆景是山石盆景中最为普遍的形式。（图4-1）峰状山石的造型，山石的形态须挺拔峻峭。主峰在高度和粗度上都必须超过其他的石料，主峰体现一种堂堂正正的威仪，客峰或前或后，呈现趋随其后的态势。在两组或三组以上的峰状山石的组织上，山峰与山峰高矮粗细都需要有一定的变化，峰与峰之间要有迎让、呼应、拱揖、顾盼，以形成一个活泼的整体，也不能各自为阵，"另立山头"。峰状山石可以结构成多种形式的盆景，例如群峰式、独峰式、双峰式、多峰式等。在群峰式的组织中，要注意山势的起伏节奏，周流环抱，参差错落；独峰式的造型往往表现的是"一峰独秀"，要做足清秀的审美内容；双峰、多峰式可以表现峡谷、溪涧等景致，要注意顾盼、呼应的关系。

峰状石盆景造型的口诀为：

高耸雄劲、山势环抱、参差错落、平中求奇。

Peak-shape style is one of the common forms of Landscape Penjing. (Figure4-1) The image of a towering and steep mountain should be presented. The major peak must be highest and thickest among the materials, embodying a magnificent impression, then the subordinate parts range by its side. If more than one group of peaks are organized, peaks should differ from each other in height and girth, and still compose a lifelike mountainous landscape. This penjing style can be further categorized as group peaks, single peak, twin peaks, multiple peaks and so on. Group peak style usually

图4-3 悬崖新雨（斧劈石；悬崖造型）Rainy Cliff(Micalex)

displays the landscape of undulating mountains encircled by water; single peak highlights its outstanding natural grace and beauty; styles of twin and multiple peak can create the vivid landscape like gorges, mountain streams, etc..

二、岩状山石盆景造型
Cliff-shape Landscape Penjing

山的边缘陡壁称崖，突出陡壁之外的称悬崖，其势险峻的称岩（图4-2、图4-3）。岩状山石因山头悬垂，重心于悬垂的方向有"塌"、"倒"、"落"的感觉，所以能造成具有险峻的动感。岩状山石造型的重心往往是成败的关键：稳而不险不生动；险而不稳则不安。因此，在险中求稳，必须注意下面几个方面：①山头悬空的大小需要适宜：头大脚小险而不稳、头小脚大稳而不险；②山腰的造型臃肿不得、瘦弱不得、对称不得：臃肿在视觉上感觉不险，瘦弱则感觉撑不住，视觉不安，对称显得呆板；③坡脚要向山头悬垂的方向延伸，上下呼应，才能寓险于稳，以达到均衡的视觉效果。

岩状山石盆景造型的口诀为：险峻幽深、钟乳悬垂、藏露有法、动静结合。

Cliff means a steep high face of rock. (Figure 4-2) Since the mountaintop is hanging down, conveying the feeling of collapse and fall, the cliff-shape landscape looks precipitous. However, steady elements are still necessary to ease the unsettled parts. Here are some points for attention:

A. The falling part should be under control, neither too big nor too small;

B. The mountainside should not be too thick, too thin or symmetric;

C. The piedmont should extend toward the pendent mountaintop to attain visual balance.

三、岭状山石盆景造型
Range-shape Landscape Penjing

"横看成岭侧成峰"，峰峦起伏连绵延续、高大的山脉称之为岭（图4-4）。岭状石的加工要点在于山之

图 4-4 岭状山石盆景造型 Range-shape Landscape Penjing

脉络上，山与山的互相连接、起伏变化，构成山系，峰脊线决定了山山连接基本形象，俗称"龙脉"。起伏之中的"起"是上升，"伏"是下降，山脉在连绵不断的上升和下降中得以延伸，这是岭状山石盆景造型的关键和核心。山脉的上升和重叠是"聚"是"收"，山势的下降和延伸是"散"是"放"，岭状山石的造型就是运用聚与散、收与放、起与伏、升与降、驰与张对立统一的美学法则，来表现山重水复、逶迤蜿蜒的山岭的造型，画论中说得好："一收复一放，山渐开而势转；一起又一伏，山欲动而势长。"比较准确地阐释了岭状山石的造型原则。

岭状山石盆景造型的口诀为：平远清逸、去来自然、奔趋有势、动静结合。

Range means a series of rolling hills or mountains. (Figure 4-4) Special attention should be paid on the processing of the mountain chain. A series of hills join together and form a mountain system. The fundamental undulating image of the mountain range is determined by the ridge line, meeting when going up while parting when going down, reflecting the unity of opposites. Generally speaking, the range-shape penjing creates landscape surrounded by mountain ranges and girdled by winding rivers.

四、石状山石盆景的造型
Stone-shape Landscape Penjing

在大自然中有很多山石，体态或玲珑剔透，或雄浑质朴，其大型石在山野中形成景观，如黄山的"飞来石"、"金龟探海"、"猴子望太平"、"仙人晒靴"等，都属于大自然中的石状山石。在中国园林艺术中，也有许多石状山石的景点，有些观赏石也属于石状山石（图4-5、图4-6）。在盆景作品中，一些倚石式或一些附石的作品选用的是石状山石，还有一类树石盆景和山石盆景的点石和近景的一些地方，以石状山石来造型作品。

石状山石盆景的造型口诀为：体态玲珑、瘦漏透皱、奇丑不陋、险稳相依。

There are various stones in nature. Some of them are ingenious and exquisite, and some are pristine and unadorned. Many large-scale ones in the mountains and fields evolve natural rock scenery. Ornamental stones are also widely seen in Chinese horticulture as scenic spots. (Figures 4-5,4-6) Besides, stone-shape rocks are the common components in penjing creations.

五、组合山石盆景的造型
Combination Landscape Penjing

组合山石盆景有两种制作方法：①用水泥将山石直接胶合在盆中；②胶合山石时在盆中垫上纸或薄膜。前者适合大型的景点或过高过险的山石造型；后者山石造型后可以搬移，任意结构，达到多景广设、多式组合，获得组合多变的艺术效果。

山石盆景要达到组合多变，必须遵循几个原则：

（1）石质相同、石色相近、石纹相似、脉理相通。

（2）加工自然、造型完整、步移景换、四面观景。就是不仅正面造型完整，反面和侧面都要完整，具有欣赏价值。

（3）在组合中要注意近景、中景、远景三者之间的变化。画论中说："远树无枝，远山无皱。"山石的纹理粗细、清晰和模糊、厚与薄等因素，都能起到表现空间关系的艺术效果。

（4）山石在造型时要势取相背，在组合时自然。一般来说，走势向左的山石置入盆的右方，右边聚左边散；反之亦然，前面聚后面散，这也是空间结构的表现方法。

图 4-5 石状山石、树木组合盆景 Stone-shape tree-rock Penjing

（5）在组合时，以高远法求高山仰视以表现雄伟的气势，以深远来表现深山雄浑的深沉厚度，以平远法来处理登高远眺、远山相御的广阔无垠和浩渺。三法并用则需单体局部的完整，仰视、平视和俯视都能够经得起推敲，都能够耐看。

组合山石盆景的造型口诀为：多景广设、多式组合、三景一体、三远并用、以峰为主、"十味调和"、乱中求整、组合多变（图4—7、图4—8）。

There are two methods to produce combination landscape penjing:

A. Glue the rocks and container together with cement, which is suitable for large-sized or excessively high landscape;

B. Put paper or film in the container, by which the rocks can be moved and dislocated, forming landscape full of variety.

Here are some rules for making various combinations:

A. Use stones and rocks with same texture, similar color and line, and linking grain;

B. Maintain natural appearance on every side after processing;

C. Pay attention to the spatial relationship between views with different visual distance;

D. Generally speaking, the rock leaning to the left should be placed in the right side of the container, and vice versa; the prominent part should face to the front;

E. The landscape should be complete and convey visual beauty from different aspects: lengthwise grandeur, transverse expanse and distant perspective.

Illustrations of combination landscape penjing are shown in Figures 4-7, 4-8.

图4-6 石状山石盆景造型 Stone-shape Landscape Penjing

图4-7 组合山石、多景广设，以岩状山石为主组合成的山石盆景 Combination Landscape Penjing, emphasizing on cliff-shape stone

图4-8 "十味调合" 以峰为主、多式组合：峰、岭、岩、洞、台、坡、峡、涧、壑、矶等并用。Combination Landscape Penjing, emphasizing on peak

第二节
山石盆景的作品展示
Works of Landscape Penjing

- ◎ 作品名称：千峰竞秀
- ◎ 石种：钟乳石
- ◎ 盆长：120cm
- ◎ 心得：这是笔者20多年前创作的山石盆景。作品表现鄂南山区千里冰封而又千峰竞秀的自然景观，表现了祖国山河如此多娇的意境。缺点是配植不够，右边配峰和坡脚不精细。

- ◎ **Title of Work:** Contending Peaks
- ◎ **Stone kind:** Stalactite
- ◎ **Basin length:** 100cm
- ◎ **Creation experience:** It is a work of landscape penjing which was created more than 20 years ago, showing the natural view of contending peaks locked in snow and ice.

- ◎ 作品名称：别有洞天
- ◎ 树种：博兰
- ◎ 石种：斧劈石
- ◎ 盆长：110cm
- ◎ 心得：作品山石造型"钟乳悬垂、险峻幽深"。树木配植多而不乱。缺点是靠盆左边一颗高树与整体不协调，需降低高度。

- ◎ **Title of Work:** Altogether Different World
- ◎ **Tree species:** Poilaniella fragilis
- ◎ **Stone kind:** Micalex
- ◎ **Basin length:** 110cm
- ◎ **Creation experience:** This work shows a precipitous scene. Whereas, the tree on the left is redundant.

◎ 作品名称："我爱五指山，我爱万泉河"
◎ 石种：沙碛石
◎ 盆长：130cm
◎ 心得：这是一件表现海南风光的作品。笔者在海南生活了20年，对这片热土怀有深厚情感。作品表现五指山和万泉河之新貌 山岩起伏，树木葱茏，一片和谐美丽、自然如花之景观。缺点是盆面空间小了一点，植物配置稍偏大。

◎ Title of Work: Five Fingers Mountain & River of Ten Thousand Springs
◎ Stone kind: Sand stone
◎ Basin length: 130cm
◎ Creation experience: It is a work that shows the scenic beauty of Hainan, a verdant island with undulate mountains. It will be better if the basin is a little bit larger.

◎ 作品名称：春山新雨
◎ 树种：博兰
◎ 石种：斧劈石
◎ 盆长：126cm
◎ 心得：作品主次分明，近、中、远景布局合理，整个画面清新明快。缺点是左边主景石稍显笨重，盆面空间还不够。

◎ Title of Work: Splendor of Spring
◎ Tree species: Poilaniella fragilis
◎ Stone kind: Micalex
◎ Basin length: 126cm
◎ Creation experience: Making a distinction between the important and the lesser one, this work is noted for its freshness and vividness. Nevertheless, the major stone on the left is a little bit cumbersome.

◎ 作品名称：峥嵘
◎ 石种：海母石
◎ 盆长：100cm
◎ 心得：作品利用海母石天然外貌（山形）进行加工制作，给人以山崖叠嶂、绵延起伏的景致。缺点是配植不够精细。

◎ **Title of Work:** Mountains Towering High
◎ **Stone kind:** Poilaniella fragilis
◎ **Basin length:** 100cm
◎ **Creation experience:** aking advantages of the original appearance of the Corallite, this work shows the rising and falling in long gentle slopes. However, the plants do not match well with the stones.

- 作品名称：一江春水向东流
- 石种：斧劈石
- 盆长：130cm
- 心得：作品以左边一块大山石为主体。其他均为陪衬或点缀。作品简洁明快，以少胜多，意境深远。

- Title of Work: A River of Spring Water
- Stone kind: Micalex
- Basin length: 130cm
- Creation experience: It is a neat and sprightly work that manifests poetic imagery.

◎ 作品名称：南江渔歌
◎ 石种：斧劈石
◎ 盆长：100cm
◎ 心得：作品为主宾式山水盆景造型，左边山石为主体，峰崖结合，渔歌唱晚，体现海南南渡江上游地区人民生活日新月异之精神风貌。缺点是远景山石偏高，意境不深。

◎ **Title of Work:** Fisherman's Song
◎ **Stone kind:** Micalex
◎ **Basin length:** 108cm
◎ **Creation experience:** This work combines both peak and cliff-shape stones, showing the life of the people who live in the area of Nandu River.

- 作品名称：南国洞天
- 石种：火山石
- 盆长：120cm
- 心得：作品充分利用山石天然和奇特之优点，见机取势，灵活多变；山石右边一组博兰林压住重心，起到了稳定作用。缺点是主峰稍偏中，应向右移，局部坡脚有欠缺。

- Title of Work: Flood Dragon in the Sea
- Tree species: Micalex
- Basin length: 120cm
- Creation experience: It is an oversize work of landscape penjing. The main feature on the right looks just like a flood dragon in the sea, and the stones on the left chime in from the other end. Nevertheless, the toe of slope needs to be improved.

- ◎ 作品名称：天上人间
- ◎ 树种：博兰、香兰、六月雪
- ◎ 石种：钟乳石、云盆
- ◎ 盆长：108cm
- ◎ 心得：作品为"盆景法"造型。用盆以一块天然云石加工成盆，再在盆上植树。山石造型高耸入云，植物配搭参差有致，整个画面和谐天成、犹如仙境，故名为"天上人间"。缺点是再加强一些透视感可能会更好。

- ◎ Title of Work: Heaven on Earth
- ◎ Tree species: Poilaniella fragilis, Chinese cymbidium, Serissa Japonica
- ◎ Stone kind: Stalactite, marble
- ◎ Basin length: 108cm
- ◎ Creation experience: Trees in this work are planted in a basin that processed from a marble stone. The mountain rocks tower over the well-spaced trees, making a harmonious picture like a heaven on earth.

◎ 作品名称：徽州人家（笔者应邀为安徽鲍家花园盆景园主景点而作）
◎ 树种：罗汉松、五针松、金钱松、榆、六月雪等
◎ 石种：芦管石、沙碛石
◎ 盆长：880cm
◎ 心得：这是一件特大型组合山石盆景。作品借鉴山水画理，融"三法"（高远法、深远法、平远法）一体，"三景"（近、中、远景）并用。多式组合，气势恢宏。

◎ Title of Work: Heaven on Earth
◎ Tree species: Podocarpus macrophyllus, five-leaved pine, golden larch, Ulmus Pumila, Serissa Japonica and so on
◎ Stone kind: Reed tube stone : Sand stone
◎ Basin length: 880cm
◎ Creation experience: It is an oversize work of combination landscape penjing, conveying visual beauty from different aspects.

◎ 作品名称：徽州人家（续）
◎ Title of Work: Heaven on Earth

第三节
山石盆景的创作演示
Creation demonstration of Landscape Penjing

一、山石盆景创作演示之一
Demonstration of Landscape Penjing(1)

2012年4月13日，笔者应邀在海南花卉大世界锦园盆景教研室做了一场山石盆景专场示范表演。

1. 表演所用大理石盆，长110cm，宽60cm；
2. 表演石料为斧劈石；
3、4. 切割和加工石料；
5、6、7、8. 摆放山石，调整和审视角度；
9. 调和水泥（用高标号水泥与107#胶水进行调和）；
10、11. 焊接山石（将报纸铺放在盆上，以防水泥与盆胶粘在一块，水泥凝固后再把报纸撕掉）；
12. 山石中间留有种植场，选择小型博兰树种植其内；
13. 作品初步完成后的效果；
14. 在山石上点缀珍珠草等小植物；
15. 清洗盆面；
16. 作品完成后的全景效果。

Figure1. A marmoreal basin with a length of 110cm and a width of 60cm;
Figure2. Stone kind:Micalex;
Figure3,4. Cut and process the stones;
Figure5 to 8. Arrange the stones;
Figure9. Mix high quality cement and 107 glue;
Figure10,11. Fix the stones;
Figure12. Plant in the little trees;
Figure13. Half-finished work;
Figure14. Add some little plants on the stones;
Figure15. Clean the surface of the basin;
Figure16. Finished work.

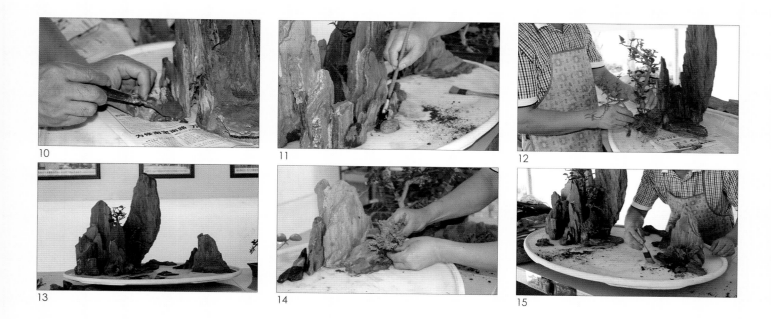

二、山石盆景创作演示之二
Demonstration of Landscape Penjing(2)

2012年4月15日，笔者在其盆景艺术中心做了一盆山石盆景创作演示。

1. 表演材料为海母石；
2. 表演用盆为大理石盆，长116cm，宽62cm；
3. 切割和加工石料；
4. 进行初步摆放；
5. 调和水泥，准备焊接石料；
6. 焊接前盆底铺上报纸，以防石料与盆粘接；
7～10. 用水泥焊接山石，注意事项：①焊接要牢固；②石料（石种）要统一；③山石与山石之间纹理脉络要相通；
11、12. 配植已培植好的小型博兰树；
13. 对所配树木进行修剪与整型；
14、15. 配植完成后，清理盆面，并将盆底报纸（山石部分除外）撕掉；
16. 作品完成后的全景效果。

Figure1. Stone kind : Corallite;
Figure2. A marmoreal basin with a length of 116cm and a width of 62cm;
Figure3. Cut and process the stones;
Figure4. Arrange the stones;
Figure5. Mix the cement;
Figure6 to 10. Fix the stones;
Figure11,12. Plant in little trees;
Figure13. Prune the trees;
Figure14,15. Clean the surface of the basin;
Figure16. Finished work.

12

13

14

15

16

三、山石盆景制作要领

石色相近石质同，石纹相似脉理通。
腹有诗画书卷气，胸中自有山万重。
高低错落立主宾，平中求奇势恢宏。
悬垂得体幽意深，险中见稳静寓动。
聚散合理清逸美，一起一伏往来中。
瘦漏透皱丑不陋，加工自然合理用。
乱中求整形分类，多式组合二神融。

第五章
树石盆景创作探索
A Survey of Tree-Rock Penjing

第一节
树石盆景主要创作形式探讨
Main Styling Forms of Tree-Rock Penjing

　　树石盆景在中国盆景的历史中得到了充分的发挥和表现，从唐朝章怀太子墓中的壁画侍女手捧有树有石的盆景，经过历代艺术家创造和发展，成为今天我们民族喜闻乐见、雅俗共赏的盆景艺术形式。树石盆景最能表现大自然的丰姿神采，突出中国盆景诗情画意的艺术特色，充分展示自然美，创造艺术美和意境美，体现作品的思想性，传自然之神和作者之神，代表着中国盆景艺术的发展方向。树石组合，从古到今，创造出多种形式的法则，我们要在传承的基础上，归纳和研究过去，目的是放眼未来，开拓创新（图5-4、图5-10）。

Tree-Rock Penjing Style, which can be tracked back to the Tang Dynasty, has a long history in China, and has become works of art that appeal to both refined and popular tastes. Conveying poetic charm, natural and artistic beauty, it represents the development course of Chinese penjing arts.

一、点石式
Ornament Style

　　点石式盆景是在盆面或土壤中点布一块或数块山石，以此表现地貌环境，或协调构图分布上的轻重，达

图5-1 点石以表现地貌、协调轻重、达到均衡 Ornament style is widely used to show the topographic feature and balance the composition

图5-2 点石式树石盆景，此法要求选用大小、重轻、主次不同之石的组合，讲求虚实、疏密、聚散的空间效果。（依据贺淦荪创意所画）Ornament style. The ornamental rocks should be varied in size and weight, and arranged in a proper way to cooperate with the trees. (Drawing from He Gansun's creative conception)

图5-3 倚石式树石盆景，以石状石为造型特点，树依石而安，多用于表现清逸的庭园小品。（依据贺淦荪创意所画）Leaning style. Trees and rocks lean against each other, bringing out the best of an elegant and refined work of art. (Drawing from He Gansun's creative conception)

图 5-4 松涛情怀（火棘、千层石）
Feelings for Songtao(Pyracantha Fortuneana . Laminated Rock)

到均衡，增添自然野趣或弥补某些树木造型上的不足，运用得比较普遍和广泛。此法最适用古老的树桩（图5-1、图5-2）。点缀的山石，要求选用大小不一、轻重有别、主次不同的山石来组合。在制作点石式树石盆景时，要讲求虚实变化、疏密有致、聚散合理的空间效果。在点石时还应当注意树与石、石与石之间的关系，注意树形和石形的变化和统一。

Ornament style is to place one or several rocks on the surface of soil in the container, which is widely used to show the topographic feature, balance the composition, impart natural interest or make up the deficiency of trees, especially suitable for the old stumps. (Figures 5-1 to 5-2) The ornamental rocks should be varied in size and weight, and arranged in a proper way to cooperate with the trees.

二、倚石式
Leaning Style

倚石式树石盆景最适合表现单体清逸的庭园小品，要求树木与"石状石"的和谐构成（图5-3）。树倚石而安，石倚树而稳，树石依偎相得益彰。制作倚石式盆景除树石相依相靠之外，还可以将树干或树根嵌入石缝和石穴之中，或树干抱石、绕石，使树与石和谐生成为一个整体，达到浑然天成的艺术效果。

Leaning style is an elegant and refined work of art, which is composed by trees and stone-shape rocks (Figure 5-3) .Trees and rocks lean against each other and bring out the best. On the other hand, the trunk and root can be inlaid in the stone crack or hole, or grow around the stone, creating an integral whole.

三、坡岭式
Range Style

坡岭式树石盆景适合于直干、曲干的树木或丛林的树木，以岭状石为造型特点。用于全景的布局，布于树下为石，增添坡岭的韵味，点于水中成渚，丰富河湖溪涧等水域，置于远处为山为岭，深化空间的关系（图5-5、图5-6）。坡岭式树石盆景在造型上，需要把握全局的观念，认真处理树与树之间的主宾、争让和顾盼，石与石之间的大小、轻重、纹理、脉络、聚散等。

Featured with range-shape rocks, this style usually uses trees with upright trunk, slanting trunk or group of trees. The rock can be placed under the tree as a stone, in the water as an islet, or in the distance as a mountain. (Figures 5-5 to 5-6) The problem of priority, density, size, grain etc. should be given full consideration.

图 5-5 坡岭式树石盆景（依据贺淦荪创意所画）
Range style (Drawing from He Gansun's creative conception)

图 5-6 坡岭式树石盆景，以岭状石为造型特点，用树适合于丛林、直、斜、曲干树木 Range style. Featured with range-shape rocks, this style usually uses trees with upright trunk, slanting trunk or group of trees.

图 5-7 环抱式树石盆景，以峰状石为造型特点（依据贺淦荪创意所画）Circle style tree-rock penjing, featured with peak-shape rock. (Drawing from He Gansun's creative conception)

图 5-8 环抱式树石盆景，以石状石为造型特点 Circle style tree-rock penjing, featured with stone-shape rock

图 5-9 临崖式树石盆景，以崖状石为造型特点，图为树石丛林组合。Cliff style tree-rock penjing, featured with cliff-shape rock.

四、环抱式
Circle Style

环抱式树石盆景最适合用于表现气宇轩昂之树木的造型。在树石盆景中树与石的组织结构应该有多样的方式，或以石环抱树，或以树环抱石（图 5-7、图 5-8）。环抱式是树石结构的重要方式之一。围流环抱体现的是一种流动的美，在我国太极图中得到了最充分淋漓的发挥。在环抱的树石结构中，树石之间的相互顾盼、迎让、高低错落、彼此关照为造型的要点。

This style is a composition of trees and rocks, including trees encircled by rocks and rocks encircled by trees. (Figures 5-7 to 5-8) When water flows around, it presents a kind of dynamic beauty.

五、临崖式
Cliff Style

临崖式树石盆景以岩状山石为造型，极富动感和奇险，在与树木的组合中，最适合于用斜干、临水、悬崖的树木来配景，也可以用丛林来结构，以表现中景和远景（图 5-9、图 5-11、图 5-12）。临崖式树石盆景在造型上，因山头的垂落，产生朝下坠的动势而造成视觉上的险峻，又可以用斜干、临水、悬崖的树木增添动势和险峻，因此，在制作时要注意把握能起均衡作用的因素，例如山石的坡脚、远山、树木造型中的一些横线枝托等等，都能起到调剂和缓解下坠力坠落的作用。

This style uses cliff-shape rocks to form treacherous scenery with slanting, hanging-down or group of trees. (Figures 5-9,5-11 to 5-12) The top of cliff-shape rocks is projecting, in addition, trees are growing downward, so the precipitous image is quite outstanding. Nevertheless, elements like the foot of the rock, distant mountains, horizontal branches etc. should be used to balance the feeling of collapse and fall.

图 4-10 盼归（北京植物园收藏）（博兰、斧劈石）
Longing for the Return(Collected by Beijing Arboretum)(Poilaniella fragilis . Micalex)

六、高峰式
Peak Style

高峰式树石盆景以峰状石为造型特点，以表现雄伟高耸、大气磅礴之势（图5—13、图5—14）。在高峰式的树石结构之中，直干的树木能增添高耸之势；斜干的树木有变化之功；横倚、悬崖的树木能添峰高之险。高峰式树石盆景的山石造型，石形要挺拔峻峭、大小高低错落，要克服布局主次不分；选择树木要把握一个度，要依据立意的表达去挑选达意之材，一定要注意树石盆景容易犯的树大"盆"小、上重下轻、无真实感的毛病；再则必须注意盆内各组"景"物之间的内在联系、虚实顾盼、争让呼应，或点石衔接，绝不能各自为阵，形成各个不相干的"孤岛"。

This style is featured at the mountain peak of great grandeur. (Figures 5-13 to 5-14) Upright trees in the composition strengthen the feeling of tall and erect; slanting trees add diversity; downward-growing trees make it look precipitous. The selected rocks for peak style penjing should be high and steep, and different in size and height. In addition, trees must be chosen in line with the creating conception to avoid being top-heavy and unstable foundation. The interrelationship between every element in the container can not be ignored.

图 5-11 临崖式树石盆景，峰石树木环抱式 Cliff style tree-rock penjing

图 5-12 临崖洞桥组合式盆景，用临水、悬崖树木组合，以显示动感和奇险。
Cliff style with bridge arch

图 5-13 高峰式树石盆景，以峰状石为造型特点，以表现高耸、雄伟、磅礴的气势。（依据贺淦荪创意所画）Peak style tree-rock penjing, which is featured at the mountain peak of great grandeur. (Drawing from He Gansun's creative conception)

图 5-14 高峰式树石组合，直干树木加强高耸之势、斜干树木具有变化之功、悬崖、横倚树木增添峰高之险。Peak style tree-rock penjing. Upright trees in the composition strengthen the feeling of tall and erect; slanting trees add diversity; downwardgrowing trees make it look precipitous.

第二节
树石盆景的作品展示
Works of Tree-Rock Penjing

- ◎ 作品名称：弄潮（国际金奖作品）
- ◎ 树种：博兰
- ◎ 石种：云母、海母石
- ◎ 树高：88cm
- ◎ 盆长：120cm
- ◎ 心得：作品将一株连体博兰树植入石盆中，经过20年的造型而达到完全成型。该树木悬根露爪，四面可观；枝法造型严谨，疏密相间，藏露有法；以石为盆并表现海南岛地图，正面和周边缀以白石子，以示沙滩和水意，耐人寻味。

- ◎ Title of Work: Brave Winds and Waves (International golden-prize work)
- ◎ Tree species: Poilaniella fragilis
- ◎ Stone kind: Mica, Corallite
- ◎ Tree heigh: 88cm
- ◎ Basin length: 120cm
- ◎ Creation experience: It is a work which takes 20 years to create. Part of roots is exposed and can be seen from each side. The branches are well-knit in proper density. On the other hand, the basin is made into the shape of Hainan Island, with the little white stones serve as sand beach and sea water.

- 作品名称：云崖飞渡
- 树种：榆树
- 石种：英石
- 树高：66cm
- 盆长：120cm
- 心得：巧借山石缝隙，将一榆树经多年提根种植其间，树木与山石紧"咬"不放，恰似扶孤松而盘恒。作品奔放自如、彰显狂野，动势飞扬。缺点是榆树飘枝（下垂枝）太长，应缩短一半，枝爪尚欠年功。

- Title of Work: Windswept Island
- Tree species: Ulmus Pumila
- Stone kind: Limestone from Yingde
- Tree heigh: 66cm
- Basin length: 120cm
- Creation experience: The selected tree has been skilfully planted with the stone for many years. The work looks wild and dynamic, while the descending branch should be shorter in half and the twigs still need time to be shaped.

- ◎ 作品名称：潇洒走天涯
- ◎ 树种：小叶榕
- ◎ 石种：海母石
- ◎ 树高：150cm
- ◎ 盆长：150cm
- ◎ 心得：这是一棵以海南当地小叶榕进行造型的榕树盆景。作品突出树木自然野趣，并对其大飘枝有意夸张拉长，加大动势力度，配盆及盆面坡脚处理蜿蜒迂回，变化自如，深化主题。缺点是枝法造型年功不够，土面还可适当加高。

- ◎ Title of Work: Unrestrained
- ◎ Tree species: Ficus Microcarpa
- ◎ Stone kind: Corallite
- ◎ Tree height: 150cm
- ◎ Basin length: 150cm
- ◎ Creation experience: The flowing branch is trained to grow overlength intentionally in order to show the dynamic state. However, the branches still need further training, and the soil surface should be higher.

◎ 作品名称：和谐世界
◎ 树种：福建茶
◎ 石种：海母石
◎ 树高：72cm
◎ 盆长：110cm
◎ 心得：树木奇正枯荣与共，山石争让倾盼有情，树与树、石与石、树与石之间高度吻合，达到自然和谐、神韵天成。不足之处是作品重心有些偏左，右边第一棵树木分量不够。

◎ **Title of Work:** Harmonious World
◎ **Tree species:** Carmona Microphylla
◎ **Stone kind:** Corallite
◎ **Tree height:** 72cm
◎ **Basin length:** 110cm
◎ **Creation experience:** Trees and stones combine naturally to form a harmonious world. Whereas, the trees are inclining to the left too much, and the one on the right should be bigger.

◎ 作品名称：石上疏林（常州宝盛园陈文娟收藏）
◎ 树种：博兰
◎ 石种：海母石
◎ 树高：48cm
◎ 盆长：30cm
◎ 心得：这是一件中小型树石盆景。作品利用海母石天然形态，将一株连理小博兰树种植石上，野趣自然。缺点是小博兰树造型年功不够，略显稚嫩。

◎ Title of Work: Thin Woods on Stones (Collected by Chen Wenjuan, Baosheng Garden)
◎ Tree species: Poilaniella fragilis
◎ Stone kind: Corallite
◎ Tree height: 48cm
◎ Basin length: 30cm
◎ Creation experience: It is a medium-size work of tree-rock penjing with an interlocked tree growing on a natural corallite. However, the tree still needs time to grow.

- 作品名称：一轮明月
- 树种：博兰、六月雪
- 石种：云盆
- 树高：70cm
- 盆长：110cm
- 心得：作品主树恰似弯弯月亮，衬树用以烘托主题。作品选用一云石作盆，更是深化主题思想。缺点是左边两株小树偏大，右边的一组六月雪太细小，不够分量，若将主树向右移10～15cm会比较好。

- Title of Work: Crescent Moon
- Tree species: Poilaniella fragilis, Serissa Japonica
- Stone kind: Marble
- Tree height: 70cm
- Basin length: 110cm
- Creation experience: The main tree looks like a crescent moon, and the other ones serve as a foil. It will look better if the trees on the left are smaller.

- ◎ 作品名称：万泉河畔
- ◎ 树种：博兰
- ◎ 石种：海母石
- ◎ 树高：68cm
- ◎ 盆长：150cm
- ◎ 心得：这是表现海南万泉河畔优美秀丽风光的一件树石盆景，作品自然入画，清新可人。缺点是左边树木层次不够清晰，主树分量不足。

- ◎ Title of Work: Riverside
- ◎ Tree species: Poilaniella fragilis
- ◎ Stone kind: Corallite
- ◎ Tree height: 68cm
- ◎ Basin length: 150cm
- ◎ Creation experience: This work shows the beauty of Hainan scenic spot, River of Ten Thousand Springs. It will look more natural if the trees on the left are thinner and clearer.

- 作品名称：海岛森林
- 树种：博兰
- 石种：云石
- 树高：68cm
- 盆长：136cm
- 心得：作品选用数株博兰树植入石盆中，形成一片森林景观，石盆造型宛如海岛，石盆正面及周边放置白石子，以示沙滩或水意，给人产生联想空间。缺点是云盆石再圆润一点更佳。

- Title of Work: Forest on the Island
- Tree species: Poilaniella fragilis
- Stone kind: Marble
- Tree height: 68cm
- Basin length: 136cm
- Creation experience: Several trees are planted in the basin, looking like a forest on the island. While the ornamental white stones serve as sand beach and sea water. Nevertheless, the primary trees are confused with the secondary ones.

- ◎ 作品名称：南国情怀
- ◎ 树种：小叶榕
- ◎ 石种：海母石
- ◎ 树高：86cm
- ◎ 盆长：150cm
- ◎ 心得：这是一盆较典型树石盆景，作品表达一棵百年古榕屹立于湖塘岸边，而且盘根错节，不可分离。表达作者对南国宝岛一往情深之意。缺点是树木造型（枝法）还有待加强。

- ◎ **Title of Work:** Feelings for Southland
- ◎ **Tree species:** Ficus Microcarpa
- ◎ **Stone kind:** Corallite
- ◎ **Tree height:** 86cm
- ◎ **Basin length:** 150cm
- ◎ **Creation experience:** It is a typical work of tree-rock penjing which shows an old tree with twisted roots and gnarled branches standing by a lake, expressing the feelings and thoughts for the southern island. Whereas, the training of the branches need to be improved.

- 作品名称：雪压冬林
- 树种：三角枫、六月雪
- 石种：硅石
- 树高：68cm
- 盆长：126cm
- 心得：这是笔者早年的代表作之一。作品表现鄂南山区植树造林成果及寒林雪景之美。缺点是树木近、中、远景处理不够清晰。

- **Title of Work:** Snowy Forest
- **Tree species:** Trident Maple, Serissa Japonica
- **Stone kind:** Silex
- **Tree height:** 68cm
- **Basin length:** 126cm
- **Creation experience:** It is one of Mr. Liu's representative works, showing the beauty of snowy forest in China. Whereas, the views with different visual distance are not distinguished enough.

第三节
树石盆景的创作演示
Creation demonstration of Tree-Rock Penjing

一、树石盆景创作演示之一
Demonstration of Tree-Rock Penjing (1)

2006年4月6日，笔者应中国（陈村）国际盆景赏石博览会组委会之邀，做了一次专场的树石盆景示范表演。

Invited by the committee of China International Penjing & Shangshi Expo, Mr. Liu Chuangang gave a demonstration of Tree-Rock Penjing on April 6th, 2006.

1. 表演前的树木为海南博兰；
2. 对博兰树逐株进行修剪和初步定型；
3. 表演用盆为定做的葫芦形紫砂盆，盆长260cm，最大宽度为88cm。笔者试放山石（海母石）；
4. 将铁砂网放置于盆底漏水洞中，用铁丝固定好；
5. 安放山石并用水泥胶合与固定；
6. 放置山石后放土（事先准备好的盆景专用土壤）；
7. 种植树木（种植前要对所表演树木材料进行整体构思：包括主次、大小、高低、疏密等等）；
8. 对树木及盆面进行认真审视和调整（左1为胡运骅先生作现场指导）；
9. 树木和山石搭配的初步效果；
10. 对每一个细部进行慎重处理（包括土面、配树、配石等）；
11、12. 最后将整理好的盆面细心铺植苔藓并点缀小植物；
13. 表演完成后的作品全景图，该作品命名为——"幽林听泉"。

Figure1. The selected trees (Poilaniella fragilis);
Figure2. Prune the trees one by one;
Figure3. The made-to-order pear-shaped boccaro basin (Length: 260cm, Maximum Width: 88cm). Put in the corallite;
Figure4. Lay a gauze screen on the hole of the basin and fix it;
Figure5. Fix the stone in the basin with cement;
Figure6. Put in the prepared soil;
Fiigure7. Replant the trees according to the creating conception;
Figure8. Arrange the trees under the direction of Mr. Hu Yunhua;
Figure9. Half-finished work;
Figure10. Further adjustment;
Figure11,12. Put on moss and little plants;
Figure13. Finished work: Secluded Forest.

9

10

11

12

13

二、树石盆景创作演示之二
Demonstration of Tree-Rock Penjing(2)

2008年5月23日，笔者在江苏常州宝盛园"首届中国高级盆景艺术师培训班"上做了一场树石盆景示范表演。

On May 23rd, 2008, Mr. Liu Chuangang put on a demonstration of Tree-Rock Penjing in the Baosheng Garden (Changzhou) for the 1st training class of Chinese senior bonsai artists.

1. 表演用的树木材料为海南博兰树。这些素材在不同规格塑料盆里已培养了2～3年，出枝、过渡等都比较到位；
2. 对每株博兰树进行修剪和整形（左1助手为笔者学生王礼勇）；
3. 清理根部，适当去掉一些原土；
4. 使用的汉白玉盆，长120cm、宽60cm。先在盆内安置金属丝，用以固定树木；
5. 开始定植树木，从大到小，第一棵树木为主树；
6. 调试树木的安放，修剪多余枝条；
7. 细心培土，把土壤压紧和压实；
8. 培土后清理盆面，准备安放山石；
9. 10. 加工和安放山石。山石在树石盆景中十分重要，布置合理，点石巧妙可以起到画龙点睛之效，同时亦可增加艺术意境和景观深度；
11. 山石加工和安置完后的整体效果；
12. 点缀小草和小植物；
13. 用水泥固定好每块山石。要求做到石与石之间和谐与稳固，达到虽由人作，宛如天成；
14. 山石布局的局部效果：大小相比、高下相较、争让顾盼、蜿蜒迂回；
15. 对所有树木的枝条作最后调整；
16. 赵庆泉大师（右3）对作品点评；
17. 表演完后的作品正面全景。

Figure1. The selected trees (Poilaniella fragilis) which have been planted in basins of different sizes for 2 to 3 years;
Figure2. Prune the trees with the help of Mr. Liu's student named Wang Liyong;
Figure3. Clear some soil from the roots;
Figure4. A white marble basin with wires in it to fix the trees. (Size of the basin: 120cm in length, 60cm in width);
Figure5. Replant the trees in the marble basin in order;
Figure6. Further adjustment ;
Figure7. Earth up;
Figure8. Clean the surface of the basin;
Figure9 to 10. Add the stones as a finishing touch that can bring the work to life;
Figure11. Half-finished work;
Figure12. Put on moss and little plants;
Figure13. Fix the stone in the basin with cement;
Figure14. A close shot;
Figure15. Arrange the branches;
Figure16. Mr. Zhao Qingquan makes comments on the work ;
Figure17. Finished work.

9

10

13

15

16

三、树石盆景创作演示之三
Demonstration of Tree-Rock Penjing(3)

2003年9月3日，笔者在"海口市刘传刚盆景艺术中心"做了一场树石盆景示范表演。

Mr. Liu Chuangang gave a demonstration of Tree-Rock Penjing in his art center on September 3rd, 2003.

1. 表演用盆为汉白玉盆，长150cm、宽70cm；
2. 表演树木为一棵一本多干博兰树，该树木造型10年，已接近成型（表演前对该树进行了摘叶和修饰）；
3. 表演用石为海母石；
4、5. 安放好树木后，开始试摆山石（左1为助手郑声俊、右1为学生王礼勇）；
6、7、8. 山石安放过程及效果；
9. 山石安放完后的全景效果；
10、11. 作最后修饰和整理（左1为赵晶华先生参与研究）；
12. 作品完成后的全景效果。

Figure1. A white marble basin with a length of 150cm and a width of 70cm;
Figure2. The selected tree has been shaped for 10 years, and its leaves have been picked up before the show;
Figure3. Corallite;
Figure4,5. Arrange the stone after replant the tree (L: Zheng Shengjun, R: Wang Liyong);
Figure6 to 8. Steps of arranging stones;
Figre9. Half-finished work;
Figre10,11. Further adjustment (L: Mr. Zhao Jinghua);
Figure12. Finished work.

1

2

3

4

5

6

7

8

9

10

11

12

四、树石盆景制作要领

树石相依动为魂，雄秀结合相辅成。
石因树活石则生，树因石存树则灵。
表现地貌显环境，协调布局重与轻。
树石相倚得安稳，和谐树形与石形。
主宾顾盼相呼应，切莫各自闹离分。
彼此关照为一体，树石浑然似天成。
自然清新意境美，饱含画意与诗情。

第六章
小型与微型盆景创作探索
A Survey of Small and Miniature Penjing

第一节 小型与微型盆景主要创作形式探讨
Main Styling Forms of Small and Miniature Penjing

一、盆景的规格尺寸
Sizes of penjing

依据盆景的规格尺寸，可以将盆景分成为特大型、大型、中型、小型、微型。120～150cm 盆景为特大型盆景、80～120cm 为大型盆景、40～80cm 为中型盆景、15～40cm 为小型盆景、10～15cm 为微型盆景、小于10cm 称为掌上盆景或袖珍盆景。

本节重点讨论的是小型与微型盆景。小型与微型指的仅是尺寸，其造型当然包括树木盆景、山石盆景和树石盆景。

Works of penjing come in several sizes: oversize (120cm to 150cm), large size (80cm to 120cm), middle size (40cm to 80cm), small size (15cm to 40cm), mini size (10cm to 15cm), and pocket size (less than 10cm).

This sector mainly discusses works of small and miniature penjing, including tree, landscape and tree-rock penjing in such sizes.

二、小型与微型盆景的由来
The origin of small and miniature penjing

在中国古代，无论是唐朝章怀太子墓穴壁画中侍女捧着的盆景，还是元代流行的"些子景"，都是小型盆景。盆景体量小型化，对盆景的普及与推广起到了促进的作用。明代屠隆在《考槃余事》中

图 6-1 树石微型盆景 Miniature tree-rock penjing

写道："盆景以几案可置者为佳，其次则列之庭榭中物也"，所提倡的也是小型与微型盆景。在海外，也将小型与微型盆景称之为"迷你盆景"。这都表明了古今中外对小型与微型盆景的重视和推崇。盆景作品趋向小型化、微型化是盆景从庭院进入家庭的必然。研究探讨小型与微型盆景的创作，无论是对盆景初学者，还是对专业人员都是值得重视的课题，特别对那些寸土寸金的地

图 6-2 小型与微型盆景组合

区和作者,小型与微型盆景具有不可抗拒的诱惑。通过对小型与微型盆景的实践,去积累盆景创作方方面面的经验,对于中、大型盆景制作上的把握,以及树石盆景创作素材的积蓄等都是有益的(图6-1,图6-2)。

Whether in China or other lands, in ancient times or today, small and miniature penjing has been held in esteem. As an inexorable trend, the miniaturization is conducive for works of penjing to entering houses from courtyards. The small and miniature penjing calls for investigation, and makes valuable contributions to the training of larger-sized works(Figures 6-1,6-2).

三、小型与微型盆景的用材、造型与管理
Materials, shaping and maintaining of small and miniature penjing

1. 小型与微型盆景由于体量上的微小,小巧精致是其主要特征。因此在用材上,树木要选择枝叶细小、叶节短密、萌发力强、耐修剪的品种;山石需要选择纹理细腻、质地温润、变化丰富的石种。

树木的用材来源,采集野生苗木或人工繁殖(如播种、扦插、压条、嫁接等)。石料可以在不同石种中挑选精致小巧的石料,或在观赏石中去挑选也是不错的选择。

A.In order to train this kind of small and exquisite penjing, the selected trees should be of tiny branches, short and dense leaf segments, strong germination and easy to pruned. On the other hand, diverse fine-grained, smooth stones are the best choice for ornament.

2. 小型与微型盆景造型上的特色

小型与微型盆景在造型上要精致,但不能失去气势。正如贺淦荪先生所说:"小中见大、缩龙成寸,而不是缩龙成蛇、缩龙成蚯蚓,而是要缩成能活跃于渊、兴风作浪、呼风唤雨的龙的形象。"故在造型上,体量微小而不失古树名木、山川形胜的风采,精致洗练而能充分展示形体结构之美感。因为精致,一枝一叶都关系着比例上的形态与气势,所以需要认真、慎重、反复推敲。例如微型树木,有时一剪下去,如"牵一发而动全身",形态或者可能焕然一新,也可能后悔莫及。这是因为看似小小一枝,它在结构比例上可能占有关键的位置。

B.Features of small and miniature penjing in shaping

Although this kind of penjing is small in size, it should still manifest a scene of trees and hills in well-connected structure. Since every branch and leaf may affect the shape as a whole, the arrangement needs repeated deliberation.

3. 小型与微型盆景的管理

小型与微型盆景在造型上的精致不仅需要制作技术上的细腻,也体现在养护的年功上。小型与微型树木造型及养护要有阶段性,需要放长时得放长,当促使密集时得修剪摘叶。骨架枝放长时,使用蟠扎调整树枝的走向与角度,或控制顶部,任其长粗长野,直到枝与干的比例达到满意的程度。当骨架枝确立后,促使枝叶密集

便上升成主要矛盾，视其树木的生长情况，当修剪则修剪，当摘叶则摘叶，每年一至三次不等，只有年复一年地反复地修剪和摘叶，枝叶方能达到密集丰满、形态精致。

C. Maintaining of small and miniature penjing

The training and maintaining of small and miniature penjing is a staged process. When the frame branches are long enough, wiring is necessary to control the growing direction until the trunk and branches are in satisfactory proportion. After radicating the frame branches, repeated trimming and leaf-picking becomes the major training work to make it dense.

Little soil in the basin may cause the evaporation of water easily. Therefore, watering should not be neglected in the maintaining of small and miniature penjing. Whereas, different plants have different water demands. Generally speaking, in spring and autumn, watering is necessary in every morning and evening, and more in summer and less in winter.

Organic fertilizer is the best choice for the fertilization of small and miniature penjing. While the concentration needs to be controlled within measure if chemical fertilizer is used. Furthermore, pruning root and changing soil are also indispensable maintaining steps for this kind of penjing.

4. 小型与微型盆景的养护

小型与微型盆景的树木因容土量少，水分容易蒸发，补充水分是不可忽视的问题。但是，要注意依据不同植物对需水量的要求来提供水分。一般来说，春秋季节每天早晚各浇水一次，盛夏严冬视实际情况有所增减。可采用沙床润水的方法减缓浇水的次数，对树木植物的生长也非常有利。

小型与微型盆景的施肥以有机肥料为主，宜薄肥勤施。如用化肥，一定掌握好浓度，否则可能造成前功尽弃。

小型与微型盆景的树木换土一般一至二年，应该根据树木长势而定。长势健壮的树木每年都得进行。因其容土量少，树根容易长满盆内，剪除三分之一或二分之一的根梢，换进肥沃土壤是一项不可缺失的工作。

5. 小型与微型盆景的病虫害防治

小型与微型盆景树木常见虫害有蚜虫、红蜘蛛、介壳虫等，病害常见的有根腐病、煤烟病、白粉病等。其原因是过潮湿、不通风、闷热并受真菌、蚜虫、红蜘蛛、介壳虫的侵蚀所致。防治的方法：改善环境、注意光照和通风、疏松土壤，针对具体的病害、虫害用适量农药。

D. The prevention and control of plant diseases and elimination of pests

Humidification and stuffiness may cause plant diseases and

图 6-3 两点式小型盆景 Two-point form small penjing

图 6-4 三点式小型盆景 Three-point form small penjing

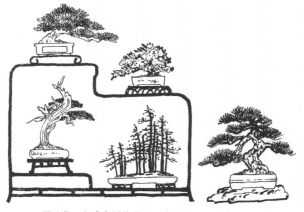

图 6-5 五点式小型盆景 Five-point form small penjing

insect pets in small and miniature penjing, such as root rot, dark mildew, powdery mildew, aphid, spider mite, coccid and so on.

Methods of prevention and control: to improve the environment, pay attention to illumination and aeration, loosen the soil and use appropriate amount of pesticide according to specific diseases.

四、小型与微型盆景的陈设
Display of small and miniature penjing

小型盆景的陈设，依据作品的大小来设置，有单体式、两点式、三点式、五点式等，微型盆景的陈设多采用几架组合（图6-3至图6-8）。在布置和组合小型与微型盆景的时候，有几点值得我们认真处理。

（1）小型与微型盆景的组合，要么选用同一的树种、石种，要么每盆的树种、石种不一样。

（2）小型与微型树木盆景配盆形状、色彩、大小要在变化中统一，配座也需要有变化。

（3）在组合结构中每件作品的样式要有变化，不要都是一种形式（专题除外）。

（4）在几架内，盆景与格内空间要大小适宜，既不可堵塞过满，又不要空旷有余。

（5）几架划分的空格，看似独立的空间，就整体布局来说，它们又是相互联系的。因此在组合布置时，依据构思的需要，突出主题，把握住顾盼、呼应、争让、虚实等对立统一的原则，使之结构成为一个和谐的整体。

俗话说："麻雀虽小，肝胆俱全。"小型与微型盆景的实践，对我们享受盆景艺术的乐趣和提高盆景技艺是大有作为的。

The display of small and miniature penjing should be in accordance with the sizes of works, including simplex form, two-point form, three-point form, five-point form, etc. Stands and Shelves are commonly used for the display. (Figure6-3 to 6-8) Here are several essentials for the combination and display of small and miniature penjing:

A. The selected trees and stones in the combination should be the same kinds, or all in various kinds.

B. The basins and stands should be different to each other while still in tuneful unification.

C. The works of penjing should be in different styles. (Except for a special subject)

D. There should be proper space in the frames of the shelf.

E. The frames of the shelf should combine together appropriately as a whole.

图 6-6 山石微型盆景 Miniature landscape penjing

图 6-7 "迷你"盆景组合 Combination of miniature penjing

图 6-8 综合组构微型盆景 Multiple composition of miniature penjing

第二节
小型和微型盆景的作品展示
Tree-and-Rock Penjing of exhibition

- 作品名称：一览众山小
- 树种：博兰
- 石种：英石
- 树高：15cm
- 盆长：10cm

- Title of Work: Overlook the Others
- Tree species: Poilaniella fragilis
- Stone kind: Limestone from Yingde
- Tree heigh: 15cm
- Basin length: 10cm

- 作品名称：别有洞天
- 石种：英石
- 盆长：15cm

- Title of Work: Amazing Caves
- Stone kind: Limestone from Yingde
- Basin length: 15cm

- 作品名称：五指山新雨
- 石种：英石
- 盆长：18cm

- Title of Work: Rain in Five Fingers Mountain
- Stone kind: Limestone from Yingde
- Basin length: 18cm

- 作品名称：龙腾
- 树种：水梅
- 树高：16cm
- 盆长：10cm

- Title of Work: Rising Dragon
- Tree species: Wrightia Religiosa
- Tree heigh: 16cm
- Basin length: 10cm

- 作品名称：明月千秋
- 树种：博兰
- 石种：火山石
- 树高：36cm
- 盆长：15cm

- Title of Work: Aged Moon
- Tree species: Poilaniella fragilis
- Stone kind: Volcanic rock
- Tree heigh: 36cm
- Basin length: 15cm

- 作品名称：舒广袖
- 树种：香兰
- 树高：36cm
- 盆长：12cm

- Title of Work: Flick Sleeves
- Tree species: Chinese cymbidium
- Tree heigh: 36cm
- Basin length: 12cm

- 作品名称：独秀
- 树种：水梅
- 树高：16cm
- 盆长：12cm

- Title of Work: Solitary Beauty
- Tree species: Wrightia Religiosa
- Tree heigh: 16cm
- Basin length: 12cm

◎ 作品名称：小品石林
◎ 石种：千层石
◎ 盆长：20cm

◎ Title of Work: Stone Forest
◎ Stone kind: Poilaniella fragilis
◎ Basin length: 20cm

◎ 作品名称：有志不在年高
◎ 树种：博兰
◎ 石种：海母石
◎ 树高：15cm
◎ 盆长：20cm

◎ Title of Work: Ambitious Youth
◎ Tree species: Poilaniella fragilis
◎ Stone kind: Corallite
◎ Tree heigh: 15cm
◎ Basin length: 20cm

- 作品名称：海岛风光
- 石种：海母石
- 盆长：21cm

- Title of Work: Beauty of Island
- Stone kind: Corallite
- Basin length: 21cm

◎ 作品名称：小品组合
◎ 树种：水梅、博兰、香兰

◎ **Title of Work:** Combination of Small Works
◎ **Tree species:** Wrightia Religiosa, Poilaniella fragilis

◎ 作品名称：树石小品集景
◎ **Title of Work:** Collection of Small Tree-rock Works

第三节
小型和微型盆景的创作演示
Creation demonstration of Small and Miniature Penjing

一、小型盆景创作演示
Demonstration of small-size penjing

2012年4月16日，笔者在海口市锦园做了一场小型树木盆景——雨林式盆景造型示范表演。

Mr. Liu Chuangang put on a demonstration of small-size penjing (rainforest style) on April 16th, 2012.

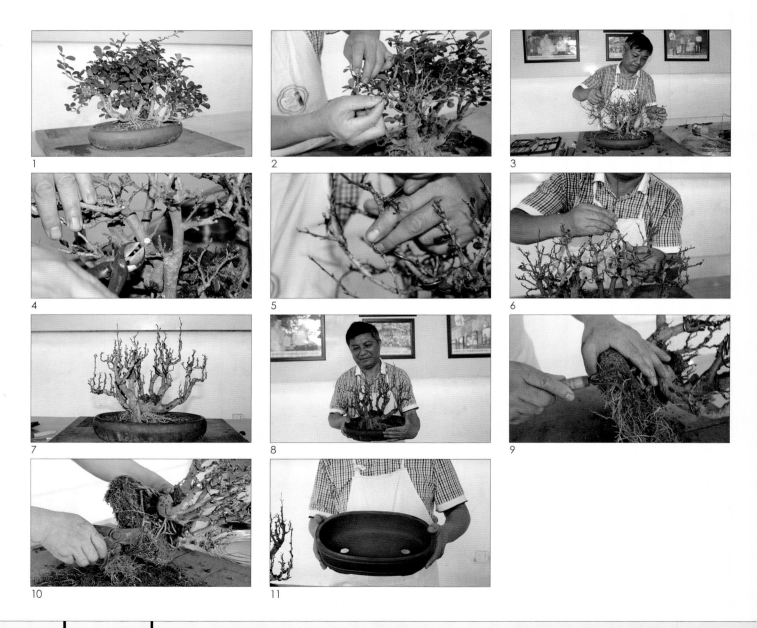

1. 表演树木为小型博兰树；
2. 开始对树木进行摘叶；
3. 摘叶完成后的效果；
4. 对树木枝条进行修剪和清枝；
5、6. 用金属丝对树木进行绑扎和造型；
7. 绑扎造型后效果；
8. 审视盆面并开始脱盆；
9、10. 将绑扎和造型后的树木从原盆中脱出，并修剪多余根系；
11. 表演用盆为椭圆形盆，长22cm，宽16cm；
12. 上盆前先将铁纱网放置在盆洞中，用金属丝固定好；
13. 开始上盆，将造型好树木放入盆中，填好营养土；
14. 细心进行培土并压实；
15. 加工石料（火山石）进行点缀；
16. 上好培植土后摆放山石并调整、审视角度；
17. 细心铺植苔藓和种植小植物；
18. 作品完成后的全景效果。

Figure1. The selected tree (Poilaniella fragilis);
Figure2. Pick the leaves;
Figure3. Tree without leaves;
Figure4. Prune the branches;
Figure5,6. Wire the branches ;
Figure7. Finish wiring;
Figure8 to 10. Get the tree out of the original basin and prune the roots;
Figure11. An elliptic basin with a length of 22cm and a width of 16cm;
Figure12. Lay a gauze screen on the hole of the selected basin and fix it;
Figure13. Replant the tree to the new basin and cover it with soil;
Figure14. Earth up;
Figure15. Process the volcanic rock;
Figure16. Arrange the stones;
Figure17. Add some moss and little plants;
Figure18. Finished work.

12

13

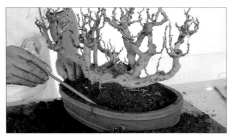
14

15

16

17

18

二、微型盆景创作演示
Demonstration of miniature penjing

2012年4月17日，笔者在其锦园教研室做了一场微型盆景创作示范表演。

1. 表演树木（博兰）及工具等；
2. 笔者表演前认真审视；3. 摘叶；
4. 摘叶后，再次观察和研究，决定做一盆微型风动式盆景；
5. 修枝，去掉多余的枝条；6. 对小博兰树进行绑扎造型；
7. 绑扎完后再次修剪和清理枝条；
8. 对修理后的枝条切面涂上愈合剂；
9. 观察造型后的效果；10. 开始脱盆；
11. 博兰树脱盆后开始梳理和修剪多余的根系；
12. 将铁纱网固定于盆底中；13. 上盆前放好营养土；
14. 安放好树木并细心把盆土压实；
15、16. 点缀山石，铺植苔藓；
17. 作品完成后的展示；
18. 作品完成后的全景效果。

Mr. Liu Chuangang gave a demonstration of miniature penjing on April 17th, 2012.

Figure1. The selected tree (Poilaniella fragilis) and implements;
Figure2. Consider how to create the work;
Figure3. Pick the leaves;
Figure4. Decide to create a windswept penjing;
Figure5. Prune the branches;
Figure6. Wire the branches;
Figure7. Prune the branches again;
Figure8. Put the consolidant on the section of the branches;
Figure9. Half-finished work;
Figure10. Get the tree out of the original basin;
Figure11. Prune the roots;
Figure12. Lay a gauze screen on the hole of the selected basin;
Figure13. Put in the soil;
Figure14. Earth up;
Figure15,16. Add some stones and moss;
Figure17,18. Finished work.

1

2

3

4

5

6

7

8

9

10

11

12

13

14

15

16

18

17

三、小型和微型盆景制作要领

小型微型基本功，中型大型一脉通。
水土精心细照料，把握采光与通风。
沙床浸水省心力，淡肥勤施切勿浓。
放长修剪分阶段，造型精致与玲珑。
依据体量来陈列，变化统一几架中。
顾盼生情成整体，各式组合要不同。
古木山川掌握内，小中见大得寸龙。

艺术交流篇
Arts Exchange

◎ 第七章　国际交流
International Exchange

◎ 第八章　国内交流
Domestic Exchange

◎ 第九章　海南花卉大世界建设
Construction of Hainan Flower World

第七章
国际交流
International Exchange

一、欧洲"三国行"纪实
Trip in Europe

2011年3月28日至4月9日,笔者与WBFF世界盆景友好联盟胡运骅主席、广州盆景协会黄就伟副会长等一行四人应邀赴欧洲荷兰、西班牙、意大利参加国际盆景交流活动,中途在西班牙与应邀参加此次活动的赵庆泉大师等人会合。我们先后参观了4家大型专业盆景公司,参加了一个国际盆景展览,进行了5场盆景示范表演活动。

● 荷兰之行令人震撼

在荷兰,我们参观了两家盆景专业公司。第一家是Lodder公司。该公司有三点令人甚为震撼。一是规模宏大;二是科学管理;三是专业化程度高。

到了荷兰,不能不看郁金香。我们参观了世界上最著名的郁金香公园——哥根霍夫公园,其感觉同样十分震撼。

● 西班牙之行领略创新

在西班牙,我们参观了MISTRAL公司。该公司最大特点是在盆景文化的推广、传播和运用方面善于创新。

MISTRAL公司是由著名盆景人士劳尔先生和其搭档共同创办的,公司员工150人左右,是西班牙最大最知名的盆景公司。

该公司非常注重交流,每两年举办一次大型国际展览会,出版盆景专业刊物,而且质量优、档次高。本次活动之前,专门刊发了赵庆泉大师和笔者两个专版介绍(分别为西班牙文和法文),对推动国际盆景文化的交

Lodder先生与笔者在荷兰郁金香公园合影

笔者和胡运骅主席、Lodder先生、黄就伟先生在荷兰盆景公司合影

笔者(左1)和胡运骅主席(右2)、黄就伟先生(右1)与克里斯皮(中)父女合影

笔者和赵庆泉大师（左）在西班牙同台创作示范表演

笔者在西班牙进行盆景创作示范表演

流起到了推波助澜的作用。

该公司另一创新之举就是盆景艺术的交流互动与普及。

展览会上的盆景示范表演是重头戏，本人有幸做了两场演示，赵庆泉大师和黄就伟先生分别做了两场和一场演示，胡运骅先生也上台演讲。这是中国盆景人首次在西班牙示范表演及演讲。我注意到我们的表演引起了比较强烈的反响。

● 意大利之行欣赏品位

在意大利，我们拜访和参观了克里斯皮盆景公司。这是一家非常讲究品位的盆景企业，该公司盆景园的建设很注重艺术表现和文化内涵。他们善于把中、日、欧盆景艺术的特点相结合。克里斯皮父子及家人对我们十分友好和热情，也非常欣赏和热衷中式园林，而且乐意与中国同行交流，很希望中国的盆景大师、专家参与其讲课或交流。

On April, 2011, Mr. Liu went to Holland, Spain and Italy to carry out penjing promotion and communication with President Hu Yunhua as a member of WBFF from China, and put on demonstration of Windswept Penjing creation in the Spain International Penjing Exhibition.

Lodder 公司盆景展示棚一角

二、世界盆景大会——"美洲行"纪实
The World Bonsai Convention

应WBFF世界盆景好联盟及本届世界盆景大会主席Pedro J.Morales之邀，中国盆景界代表一行及江苏省金坛市政府有关领导共十余人参加了此次盛会。笔者有幸参加了此次盛会并经历了活动的全过程，现将主要活动内容作如下介绍。

● 参访美国国家植物园

该园是美国农业部的科研和教育机构，从属于农业部研究署（ARS），也可称为一座"活的植物博物馆"，该园始建于1972年，占地46英亩，分为室内展馆和室外展馆两部份，室内馆主要陈设、介绍图片和书籍，室外馆又分日本馆、中国馆、北美馆、国际馆·热带温室，馆内收集的精美盆景及相关人造景观，充分体现了中国、日本和北美传统设计风格。

● 世界盆景大会与WBFF组织

世界盆景大会（下称"世会"）是当今世界最高规格的盆景界盛会，每隔四年主办一次，距今为止已举办六届，分别在日本、美国（两届）、韩国、德国、波多黎各等国家举行。

"WBFF"是"世界盆景友好联盟"的英文简称，其会员国遍及世界五大洲。该组织一共设9个理事，包括中国、日本、印度、亚洲区、北美区、南美区、非洲区、澳洲区、大洋洲区等。

● 成功申办下届世界盆景大会

世界盆景大会在业界素有世界盆景"奥林匹克"之称，因此，申办世会是一种崇高荣誉，但决非易事。

2009年7月8日下午3时许，世界盆景友好联盟组织以及各理事国将主办权交与了中国，胡运骅先生同时也以高票当选为第七届世界盆景友好联盟的大会主席。

● 富有实效的盆景艺术交流

世会结束后，应美国西雅图、洛杉矶、三藩市等盆景协会及亚太园艺公司总裁曹伟泰先生邀请，笔者和广州盆景协会副会长黄就伟、副秘书长林南一起于2009年7月12日从波多黎各飞往华盛顿，随后转乘美国UNITED航班飞往西雅图。

7月13日上午9点，我们来到西雅图盆景博物馆，笔者应邀在该馆演示厅做了一场风动式盆景创作示范表演，表演材料为当地杂木类半成品，曹伟泰先生当现场翻译。表演前，笔者向来自西雅图地区三个协会的盆景爱好者介绍了中国盆景的特点与将要表演的"风动式"盆景形式，并根据所提供的材料现场勾画了一个设计草图，征求大家意见，大家当场就提出了许多感兴趣的问题，笔者一一作答，现场气氛比较热烈。

7月14日上午，笔者和黄就伟、林南先生又应邀在亚太园艺公司分别做了两场盆景创作示范表演。下午拜访了当地资深盆景艺人DANIEL ROBINSON，这位老先生从艺50多年，虽已近80岁高龄，但仍精神抖擞，见到来自中国的盆景人士十分高兴，中午还宴请了我们。

我们在洛杉矶的四天时间里，受到美籍华人盆景人士刘炳衡先生热情款待。先后还拜访了车兆滕先生、郭思恩先生等5家私家花园或盆景园，并与他们进行交流。其间，笔者还应邀做了一场松树盆景创作示范表演。

7月22日下午1点左右，我们结束了在美国的

美国国家植物园园长汤姆斯（右2）、盆景园负责人（左一）与赵庆泉大师（右3）、刘传刚合影（华盛顿）

笔者向美国国家植物园园长汤姆斯赠送个人盆景作品邮票专辑（2009年华盛顿）

WBFF世界盆景联盟前主席Solita夫人（左）、国际盆景顾问辛长宝（右）和刘传刚合影（波多黎各）

全部行程，从旧金山飞往上海浦东。在飞机上笔者思绪万千：笔者第一次参加世界盆景大会、中国"申办团"申办世界盆景大会获得成功，中国也是第一次成为WBFF主席国，应该说世界盆景组织回归于中国是人心所向、大势所趋，但要真正把这项工作做好做实，全面展示盆景创始国的风范，引领世界盆景新潮流，其任务之艰辛、压力之巨大不言而喻。但笔者坚信，泱泱大国的中国盆景领导者、组织者及所有的中国盆景人会为之而不懈努力！

On July, 2009, Mr. Liu participated in the World Bonsai Convention

United States National Arboretum
The arboretum functions as a major center of botanical research, conducting wide-ranging basic and developmental research on trees, shrubs, turf, and floral plants.

WBFF
The name of this organization is World Bonsai Friendship Federation, abbreviated to "WBFF." The membership of the WBFF is comprised of regional members representing nine geographical regions of the world as follows: Africa, Australia/New Zealand, China, Europe, India, Japan, Asia-Pacific, Latin America, and North America.

Successfully bid for the next World Bonsai Convention
On July 8, 2009, China successfully bid for the World Bonsai Convention for the first time, and Mr. Hu Yunhua was elected to be the chairman.

Arts exchange
After attending the convention, Mr. Liu Chuangang continued to visit the bonsai gardens in Seattle with other representatives from China, and put on several demonstrations.

笔者在美国西雅图盆景博物馆做完示范表演后与馆负责人 Darid J.DeGroot 在作品前合影

美国知名盆景老艺人（中）与曹伟泰（左1）、林南（左2）、黄就伟（左4）、刘传刚进行交流

笔者、黄就伟、林南在美国西雅图亚太公司与曹伟泰先生和夫人合影

笔者与美国 Introduction to BONSAI 主编合影

三、日本国风展——"日本行"纪实
The Kokufu-bonsai Exhibition in Japan

正值中国农历新春佳节之际,应日本盆栽协会及WBFF世界盆景友好联盟副主席岩崎大藏先生之盛邀,中日盆景友好访问代表团(下称代表团)一行20余人,于2008年2月9日(农历正月初三)从上海出发,赴日本进行了为期一周的参观、交流与访问。

● 日本盆栽协会

日本盆栽协会创办于1934年,当时名称为"日本国风盆栽协会"。昭和40年(即1965年),改名为"日本盆栽协会"。

● 国风盆栽展

国风盆栽展,是日本最高级别的盆景展览活动。该活动由日本盆栽协会主办,日本文化厅、东京都协办。国风展每年举办一次,迄今为止,已举办了81届(日本称81回),今年是第82届。笔者在日期间曾两次赴展览会现场参观与了解。

● 东京大卖场

东京大卖场位于上野公园东京美术馆附近,距国风展大约15分钟车程。大卖场是以展示、销售、交流于一体的综合性交易市场,给笔者留下了如下四点印象:一是布置整齐规范。二是展品(商品)质量好。三是盆栽配套服务齐全。四是明码标价、以质论价。

● 木村正彦盆景园

2月12日上午,代表团来到木村正彦先生的私家盆景园。该园占地面积2亩左右,但布置得十分精致,各类盆栽作品琳琅满目。精品展示区、作品陈列区、养护棚、演示厅等设施齐全、效果均佳。

● 大宫盆栽组合园

大宫盆栽组合园(亦称大宫盆栽村),是日本国内最具规模的盆栽聚集区。组合园内包括芙蓉园、一光园、留芳园、泉松园、松涛园、藤树园、蔓青园、宽乐园、清香园、九霞园等众多私家盆栽园。

● 高砂庵盆景园

高砂庵盆景园是代表团此次日本之行参观的最后一站,也是给笔者留下印象最为深刻、最难以忘却的一站。该园占地50亩,是世界盆景友好联盟副主席岩崎大藏先生投巨资(总投资超过1亿元人民币)所建,可以算得上是日本国内乃至世界最好的盆景园之一。

一是设计颇具新颖。二是精品纷呈、造型各异。三是高砂庵主人岩崎大藏先生的人格魅力为笔者及代表团每位成员所感动。

2008年中日盆景友好代表团成员和岩崎大藏先生在其园内合影

笔者在日本木村正彦盆景园留影

日本盆栽协会会长竹山浩先生（右）与笔者亲切握手（2008年东京国风展）

日本木村正彦（右）和笔者合影（2004年）

为期一周的日本之行虽然时间短暂，但记忆却是永恒的。笔者认为，我们应虚心学习国外的先进经验，洋为中用，他为我用、取长补短、不断提高。譬如：学习日本先进的造型与养护技术、盆景工具和设备的研制与生产、精到的管理与营销理念等等。同时，我们更应该走"中国特色"之路，把握宏观大局，维护国家和民族利益，这就是要创立"中国派"！只有民族的，才是最能被别人认可与推崇的，也是最能走向世界的。

On February, 2008, Mr. Liu visited the Kokufu-bonsai Exhibition in Japan as a Chinese representative with a delegation, and carried out a one-week investigation.

Nippon Bonsai Association
Nippon Bonsai Association was started in 1934.

Kokufu-bonsai Exhibition
Kokufu-bonsai Exhibition is the biggest annual bonsai event in Japan held by Nippon Bonsai Association every year.

Tokyo Marketplace
It is a composite marketplace nearby the Tokyo Metropolitan Art Museum

Bonsai Garden of Masahiko Kimura
The delegation visited the private bonsai garden of Mr. Masahiko Kimura on 12th February. The well-equipped facilities in the garden were quite impressive.

Omiya Bonsai Village
Omiya Bonsai Village is an area consisting of a dozen bonsai gardens, such as Fuyo-en, Shoto-en, Toju-en, Mansei-en, Seiko-en, Kyuka-en and so on.

Takasago-an Bonsai Garden
In Mr. Liu's opinion, it can be regarded as the best bonsai garden in Japan, or even around the world.

岩崎大藏先生盆景园地景五针松

四、马来西亚盆景古石展赛会纪实
The International Penjing & Gushi Exhibition & Contest in Malaysia

笔者向马来西亚盆景雅石会会长戴文华（右2）赠送其书法作品（2006年）

BCI亚太盆景协会主席苏义吉先生（中）新加坡谢江水先生和刘传刚合影

马来西亚教育部长YB拿督翁诗杰向笔者颁发国际评委证书（2006年9月马来西亚）

九月的马来西亚吉打州，秋高气爽，繁花争艳，景色宜人，马亚西亚首届盆景古石展赛会在这里隆重举行。笔者有幸作为海外专家代表之一，应邀参加此次展赛盛会，并被大会推选为盆景评审会主委，亲历了活动的全过程。

此次展赛会有100件精品盆景参展，根据笔者所编写的"评比方案"，最后由评委会共评出金奖作品10件、银奖15件、铜奖15件，获奖率为40%。所有获奖作品（含其他未获奖作品）均在开幕前全部公布。其严密、规范、公正的评比程序，受到了所有参展者的拥护和一致好评。

此次展赛会就规模而言，和我国国内同类展览相比要小得多，但对展品的质量和要求、展地布局效果等方面要好于国内诸多展览会。特别是宣传工作做得好、抓得实。

展赛会开幕前后五天内，马来西亚国家及地方电视台曾多次专题报道展赛会有关情况。《中国报》、《星洲日报》、《东方日报》、《北马新闻》、《光明要闻》、《光华日报》、《南洋商报》等大马国内知名报刊分别用中英文专版或大篇幅报导展赛会情况，这在我们国内是不多见的。

此次大马之行，给笔者留下了深深的、美好的回忆……

As one of the overseas experts, Mr. Liu participated in the International Penjing & Gushi Exhibition & Contest in Malaysia and acted as a judge. Although it was a small-scale exhibition, it was quite successful, especially in the propaganda work.

火炬峰（火山石）
Torch Mount(Volcanic rock)

五、菲律宾第七届亚太盆景会议纪实
The Asia-Pacific Penjing & Shangshi Exhibition & Convention in the Philippines

为期四天的"第七届亚太盆景雅石会议暨展览会"于2003年2月27日至3月2日在菲律宾马尼拉市举行，来自亚太地区近20个国家和地区分别组团或派代表参加了此次盛会，笔者和有关方面代表一行九人参加了这次大会。

笔者作为中国盆景代表团专家成员之一，与各国和地区盆景界友人进行了广泛的接触和交流，增强了相互了解，也增进了友谊，达到了预期目的。最令人欢欣鼓舞的是，中国代表团代表北京植物园申请第八届亚太盆景赏石会议暨展览举办权获得成功。

On February, 2003, Mr. Liu Chuangang participated in the Asia-Pacific Penjing & Shangshi Exhibition & Convention in the Philippines with other 8 Chinese representatives, and successfully bid for the 8th Asia-Pacific Penjing & Shangshi Exhibition & Convention.

笔者（2排左2）在菲律宾与各国盆景界友人合影

笔者（右1）与麦培满、甘伟林、苏义吉、韦金笙等国际人士合影

甘伟林理事长（中）韦金笙副理事长（左）和刘传刚在晚宴上（2003年）

第八章
国内交流
Domestic Exchange

一、主持海南省"三届"盆景评比展览
Hainan Penjing Exhibition

从 2001 年开始，由笔者总策划并主持了海南省一至三届全省盆景评比展览，在各级政府的高度重视与关怀下，在国内外同行界的大力支持与海南盆景界同仁的共同努力下，三届盆景展均取得圆满成功。

1. 海南省第一届（凤凰花城杯）盆景评比展览

2001 年 11 月 23 日，由海南省林业局、省盆景专业委员会等单位共同举办的海南省第一届（凤凰花城杯）盆景评比展览，在海口市会展中心隆重举行。此次展览汇集了全省各地盆景精品 200 盆（件），共评出金奖 15 件，银奖 20 件，铜奖 35 件。笔者选送了 20 件作品参展，但作为主办者，笔者送展作品只参展不参评。

2. 海南省（奥林匹克杯）第二届盆景评比展览

2003 年 10 月 1 日，由海南省林业局、海口市人民政府主办、省盆景专委会和奥林匹克花园公司共同承办

中共海南省委书记、省人大主任汪啸风（左 8）及夫人王莲生（右 8）参观盆景展览后和海南盆景界同仁亲切合影（2003 年 10 月 1 日海口）

海南省党政领导及中外嘉宾出席省第二届盆景展览开幕式
前排左起：笔者、李丹、黄金城（省林业局副局长）、罗伯特（印尼）、许俊（省委常委）、洪寿祥（省政协副主席）、韩至中（省人大副主任）、林方略（副省长）、符气浩（省政协副主席）、王莲生（省委书记夫人）、梁悦美（台湾）、朱勤飞（新加坡）、周燕华（省农业厅副厅长）
后排左起：刘宗达、纪进、韦金笙、冯川建（省政协常委）、胡运骅、吴伟雄（副市长）、吴敏、陆志伟、胡世勋、许家华（建设厅副厅长）、邓子芳（省美协主席）、吴华盛（省林业局副局长）

省委副书记、海口市委书记王富玉（右2）参观盆景展并听取笔者（左1）介绍海南盆景，赵晶华（左2）、韩林畴、连明良（右1）陪同参观

的全省第二届盆景评比展，在海口市隆重举行。此展全面展示了海南盆景最高水平。省委书记、省长汪啸风、省委副书记、海口市委书记王富玉、省委常委许俊、省人大副主任韩至中、副省长林方略、省政协副主席洪寿祥、符气浩、海口市委副书记张作荣、副市长吴伟雄等省、市四套班子领导分别参观或参加了展览开幕式。参加此活动的国内外嘉宾有：胡运骅、韦金笙、陆志伟、苏义吉（台湾）、梁悦美（台湾）、朱勤飞（新加坡）、罗伯特（印尼）等。开幕式当天举行了"中外盆景专家现场表演"、"国际盆景专家胡运骅、苏义吉先生专题讲座"等系列活动。

3．"海南首届国际盆景根艺赏石邀请展"暨"海南省第三届盆景评比展览"

2010年12月31日，由海南省林业局、海口市人民政府主办，海口市林业局、海南省盆景专业委员会、海南花卉大世界共同承办的此次活动在海南花卉大世界隆重举行。共展出来自美国、日本、新加坡、马来西亚等7个国家及港、澳、台地区以及省内外参展作品共600件，其中盆景300件，根艺150件，赏石150件等。除了海南省和海口市有关党政领导和新闻媒体出席了此次活动外，参加活动的嘉宾还有：胡运骅、辛常宝、赵庆泉、谢克英、李正银、郑在权、吴成发、黄就成、黄就伟、陈文娟、袁增民、张辉明、伍恩奇等。展区内还设置了"刘传刚树石盆景专题展"，笔者所有参展作品"只参展不参评"。活动期间，举行了"中外盆景专家现场示范表演"、"海南首届盆景作品拍卖会"、"海南首届中外盆景学术论坛"等系列活动。

A. 2001-10, acted as the general superintendent of the 1st Hainan Penjing Exhibition;

B. 2003-10, acted as the secretary general of the 2nd Hainan Flower Arrangement Evaluation Exhibits;

C. 2009-12, acted as the general superintendent of the 1st Hainan International Penjing & Shangshi Exhibition and the 3rd Hainan Penjing Contest.

笔者主持评比展开幕式活动

笔者主持颁奖仪式（由赵庆泉大师向彭盛才先生颁发证书）

笔者（右1）和领导嘉宾在开幕式合影（左起：赵庆泉、江长桥、辛长宝、蒙国海、胡运骅、秦忠文、张华江、林劲、郑在权、陈荣森）

二、台湾盆景艺术活动
Art activities in Taiwan

2011年11月1日至8日，笔者应台湾休闲农业组织之邀，专程赴台湾参加了休闲农业论坛与相关活动。其间，笔者一行4人专程赴台北市天母山拜访了台湾著名盆景艺术大师梁悦美教授，梁教授对笔者一行的到访十分高兴，并予以盛情款待。笔者一行在参观她的盆景园时，梁教授在介绍盆景园的特色后，逐盆向我们介绍其创作心得与作品艺术特色，我们深受感动，可谓受益匪浅。同时笔者还就国内盆景发展趋势、台湾地区盆景发展现状以及国际盆景发展动向等议题与梁教授进行了深入探讨，达成了许多共识。

On November, 2011, invited by the organization of Taiwan agriculture, Mr. Liu took a trip to Taiwan and participated in a series of activities there.

笔者（左）和梁教授共同欣赏笔者15年前赠送的海南博兰盆景

笔者（右）和梁悦美大师共同欣赏其微型盆景艺术

笔者（右2）和夫人王四英（左2）拜访台湾大伯（右3）、伯母（右4）和妹妹（左1）及妹夫（右1）合影留念

笔者（右2）和夫人参观梁悦美大师盆景园后和梁教授及先生等合影

三、主持"世界盆景友好联盟交流中心"落户海南活动
Exchange centre of World Bonsai Friendship Federation in Hainan

2011年10月4日，WBFF"世界盆景友好联盟交流中心"在南渡江畔的海南花卉大世界园区内举行揭牌仪式。这是该中心首次落户海南。

WBFF世界盆景友好联盟主席胡运骅、WBFF前任主席、现任联盟名誉主席沙丽达·罗沙狄、美国国家植物园园长、国际顾问汤姆斯、国际顾问吴成发、梁悦美等来自世界16个国家和地区的著名盆景人士出席了当天的揭牌仪式。海南省政府秘书长徐庄出席仪式并致词，原省领导董范园和省纪委、省政协、省外事办、省林业局等有关部门负责同志也出席了仪式。在开幕式前，海南省政府领导会见了世界盆景友好联盟嘉宾一行。

The exchange centre of World Bonsai Friendship Federation (WBFF) was set up in Hainan for the first time on October, 2011.

笔者（右）主持WBFF揭牌仪式

笔者（右1）在揭牌仪式后与领导嘉宾合影

省政府秘书长徐庄（左）在花卉大世界举行的"世界盆景联盟交流中心"揭牌仪式上代表省政府致词

笔者（前排左1）同前来参加揭牌仪式的中外友人合影

四、香港岭南盆景艺术活动
The Membership Representative Conference of Hong Kong Lingnan Penjing Institute

2011年春节前夕，香港岭南盆景艺术学会第二届会员代表大会暨就职典礼在香港隆重举行。参加此次活动的人员有该学会全体会员和当地政府部门领导，笔者应邀赴港参加了此次活动。应邀参加活动的嘉宾还有：广州盆景协会会长谢荣耀、副会长黄就伟、香港盆景雅石协会主席郑在权、副主席吴成发、刘耀辉等共150余人参加。笔者和谢荣耀、郑在权等嘉宾分别致词并向学会赠送书法作品或礼品。活动开始前后，与会人员分别进行了盆景艺术探讨和交流，并现场参观了学会创办的"盆景创作园"、"香港盆栽会盆景园"、"香港莲池景观"等。

On January, 2011, Mr. Liu participated in the 2nd Membership Representative Conference of Hong Kong Lingnan Penjing Institute and carried out investigation

笔者（前右3）与黄就成会长（右5）、郑在权主席（右4）、广州谢荣耀会长（左2）、黄就伟副会长（右1）等合影

笔者在香港岭南盆景艺术学会就职典礼晚宴上致贺词（2011年香港）

笔者（中）向香港黄就成会长（右）赠送本人的书法作品"岭南盆景独秀中华"

笔者参观香港盆栽会作品园并与同仁们合影

五、广州国际盆景邀请展
Guangzhou International Penjing Invitational Exhibition

2010年11月10日至11月27日，广州国际盆景邀请展在中山纪念馆盛大开幕。本次盆景展是广州地区有史以来规模最大、水平最高的一次盆景盛会。共有来自国内广东、江苏、海南、福建、江西、浙江、广西、山东等省（区）和中国香港、澳门、台湾地区以及日本、泰国等近70个送展团体800件盆景作品参展。笔者应邀担任此次活动盆景专家评审委员会评委，出席了一场"岭南盆景高峰论坛"并担任主讲嘉宾，同时做了一场风动式盆景创作示范表演。

On June, 2010, Mr. Liu participated in the Guangzhou International Penjing Invitational Exhibition as a judge, and gave a speech and demonstration in the Penjing Summit Forum.

笔者（左1）做完示范表演后和梁悦美大师（左2）、西班牙劳尔先生（左3）、朱林辉先生合影。

六、上海"东沃杯"中国盆景精品展暨创作比赛
The Dow Cup China Penjing Invitation Exhibition & Penjing Creation Contest

笔者代表全国参展单位和个人致辞

为迎接2013年世界盆景友好联盟大会在中国召开，选拔优秀盆景展品和创作人才，中国风景园林学会花卉盆景赏石分会与上海市绿化和市容管理局于2010年10月1日至10月7日，在上海市植物园举办了"中国盆景精品展暨盆景创作比赛"，笔者应邀并代表海南选送二件作品参展（均获奖），同时参加了盆景创作比赛（获金奖）。

On October, 2010, Mr. Liu took part in the Dow Cup China Penjing Invitation Exhibition & Penjing Creation Contest and won award by his own works.

笔者（中）进行丛林树石盆景示范表演

表演结束后笔者和创作作品合影（其作品赠送上海植物园收藏）

胡乐国大师（右1）、梁悦美教授（中）和笔者在其参展作品《弄潮》旁留影

七、广东（陈村）BCI 国际盆景赏石博览会
China (Chencun) International Penjing & Shangshi Expo

由中国风景园林学会、广东省佛山市顺德区人民政府和国际盆栽协会（BCI）共同主办的"2006年中国（陈村）国际盆景赏石博览会"，于2006年4月30日至5月7日，在广东顺德陈村花卉世界成功举办，应大会组委会之邀，笔者担任此次盆景专家评审委员会评委，并做了一场丛林式树石盆景专场示范表演。期间，笔者亦与来自海内外的盆景界专家学者和友人进行了广泛的接触和交流。

此次博览会以"人与社会和谐发展"为办会宗旨，中国盆景赏石、国际盆景赏石精品作品共500件参展，国内有北京、上海、天津、广东、海南、广西、山东、江苏、浙江、江西、安徽、福建、湖北、宁夏、内蒙和中国香港、澳门、台湾地区的城市组团参展。国外有美国、荷兰、西班牙、意大利、泰国、韩国、日本、新加坡、印尼等国家组团参展。

On April, 2006, Mr. Liu was invited to the China (Chencun) International Penjing & Shangshi Expo as a judge and put on a demonstration of tree-rock penjing.

笔者在国际盆景展会上做大型树石丛林示范表演，胡运骅先生（左上）主持并作演讲（旁为翻译）2006年

左起：汤姆斯、苏义吉、岩崎大藏、笔者、胡运骅担任国际评委（2006年）

笔者和美洲国际评委合影（2006年）

前世界盆景协会主席加藤三郎（中）和朱勤飞先生（右）、笔者合影（2006年广东）

八、北京第八届亚太地区盆景赏石会议暨展览会
The 8th Asia Pacific Penjing & Shangshi Exhibition & Convention

恩师贺淦荪先生（中）与笔者和夫人合影于 2005 年北京亚太会议

傅珊仪副理事长在亚太会议上与笔者合影（2005 年北京）

笔者在第八届亚太盆景会议做演讲和创作表演（夫人为助手，2005 年北京）

2005 年 9 月 6 日至 15 日，由中国风景园林学会、北京市园林局主办，中国风景园林学会花卉盆景赏石分会、北京植物园承办的第八届亚太地区盆景赏石会议暨展览会在北京植物园隆重举行，来自美国、德国、新加坡、西班牙及中国等 14 个国家和地区组团参加，展出精品盆景 360 多盆，精品赏石 500 余件。笔者十分有幸受活动组委会之邀，担任了本次展会盆景评委，同时担任评委的还有：胡运骅、胡乐国、赵庆泉、于锡昭先生等。

2005 年 9 月 8 日，笔者应活动组委会的安排，在北京植物园大会演示厅做了一场垂枝式树木盆景创作的专场示范表演。

On September, 2005, Mr. Liu Chuangang participated in the 8th Asia Pacific Penjing & Shangshi Exhibition & Convention as a judge, and later gave a demonstration of tree penjing in the Beijing botanical garden.

九、刘传刚盆景个性化邮票首发式
Release of "Personal Stamp Sheet of Liu Chuangang Penjing"

盆景界嘉宾参加笔者盆景邮票首发式后合影留念（2002年海口）

右起王选民、鲍世骐、林鸿鑫、朱勤飞、笔者、黎坚、李嘉恒参加刘传刚盆景邮票首发式后合影（2002年海口）

2002年12月9日，经国家邮政局批准，中国集邮总公司发行的"刘传刚盆景艺术个性化邮票"在海南省海口市隆重举行。海南省邮政票品局局长陈凤杰、省林业局副局长周燕华、吴华盛、澄迈县委书记杨思涛等领导，以及国内外和海南盆景界代表鲍世骐、王选民、谢克英、林鸿鑫、陈习之、黎坚、朱勤飞（新加坡）、罗伯特·史蒂文（印度尼西亚）等100余人出席了首发式。笔者对前来参加此次活动的各级领导、各位专家和同仁致以热烈的欢迎和诚挚的谢意。活动期间，大家进行了盆景艺术探讨和交流，收到了良好的效果，同时对海南盆景艺术事业的发展起到了推波助澜的作用。

China National Philatelic Corporation issued the "Personal Stamp Sheet of Liu Chuangang Penjing" in 2002, which was composed by 16 80-fen stamps.

第九章
海南花卉大世界建设
Construction of Hainan Flower World

一、海南花卉大世界简介

海南花卉大世界，是在笔者从事盆景艺术创作的同时，与公司同仁和股东一道亲历谱写的一篇新的创业乐章。2007年，原来的海口凤凰花城被拆迁改建房地产项目。笔者因为挚爱盆景艺术及花卉园艺，不忍数百花商被房地产商占地而无落脚之地，主动担纲并联合侯立军、李恒斌先生一道发起创办了海南花卉大世界。2008年至今，为拓展该项目，又先后与知名盆景人和企业家：广东中山张华江先生及深圳陈颖昌先生等股东联袂合作，使该项目运作逐步进入良性循环。

该项目座落于风景秀丽的南渡江畔，占地300亩，是经琼山区政府立项、海口市政府批准的重点项目。之初，由笔者牵头成立海南鑫山源热带园林艺术有限公司作为发起和建设单位，笔者出任总经理及项目总负责人。经过四年的努力，现已成为海南规模最大、功能最为齐全的花卉及林产品交易市场。创建四年来，实现了"四

海南花卉大世界开园仪式（2009年1月16日）

海南花卉大世界主大门（实景）

级跳"：2007年被琼山区政府确定为"重点建设项目"；2009年被海口市政府授予"花卉产业重点示范项目单位"；2010年被农业部和海南省政府联合授予"海南现代创意农业优秀品牌企业单位"；2012年开始创建国家4A级生态旅游景区，并获省、市相关部门批复。

二、海南花卉大世界创建情况

经过艰苦的努力，至2010年年底，海南花卉大世界已建成八大功能区：分别为花卉区、盆景区、景观苗木区、科研教学区、中外名园区、综合区、花卉资材交易中心、花卉超市一条街，入驻园区的国内外花卉盆景园林企业157家，总投资已超过1.5亿元人民币。

随着海南省国际旅游岛战略的实施，笔者及公司团队并不满足于把这个花卉大世界作为海南花卉、林产品展示交易的平台和出岛出口的集散地，因而提出把海南花卉大世界打造成一个以"美丽·生态·休闲·时尚"为主题，具有经济、社会、文化、教育、游憩、保健和购物与旅游等多方面功能的休闲农业品牌景区，为海口市民和岛内外游客提供一个赏花买花和休闲观光的好去处。为此，公司董事会审时度势、果断提出：再投入资金1亿元，用2～3年的时间，将海南花卉大世界创建成以花卉文化为主题的国家4A级旅游景区。同时增设三大功能区，即"琼洲文化风情街"、"生态美食一条街"、"国际休闲农园"。以上功能区已初具雏形，预计在两年内建成。相信通过我们不懈努力，海南花卉大世界一定会为海南花卉行业健康发展起到规范和促进作用，并在海南国际旅游岛建设舞台上大放异彩！

三、海南花卉大世界倍受政府关怀

自2009年开园以来，该项目得到了各级政府及有关部门领导关怀和支持。全国政协副主席、台盟中央主席林文漪莅临视察。省政协主席于迅，省委常委、海口市委书记陈辞，副省长林方略、陈成、省政协副主席王路、王应际，海口市委副书记陈军、副市长蒙国海、丁竹、谢京等领导多次亲临指导工作。省、市林业厅、局，台盟海南省委会以及国土、规划、城建、园林等各部门领导都给予了关怀和帮助。海南花卉大世界规模初成，已成为花的海洋、树的森林、景的汇集，为海南花卉产业的健康发展，以及国际旅游岛的建设发挥着重要的窗口作用。

Covering an area of 300 mu, Hainan Flower World is located by the beautiful Nandu River. Mr. Liu Chuangang is one of the promoters of this project, and now taking up the post of general manager. Hainan Flower World has gone from strength to strength since it was established 4 years ago, and already become the largest market of flowers and forest products in Hainan.

There are altogether 157 enterprises in Hainan Flower World, making it a collecting and distributing centre of flowers and forest products. Furthermore, it is given an official to be built as a national 4A-class ecotourism scenic spot, including a cultural and custom street, an organic food court and leisure farms.

On the other hand, a series of activities have been successfully held by Hainan Flower World, such as Changing Flower Festival, Spring Festival flower fair, exhibitions of penjing, stones, tourism products and so on.

海南花卉大世界琼州文化风情街主大门

海南花卉大世界一角——笔者创办的"锦园"

四、海南花卉大世界系列活动

几年来，笔者和公司团队一道，在全体股东的积极支持下，在海南花卉大世界成功举办了一系列的重大活动，如：海南花卉大世界开园典礼、海口传统活动"府城换花节"、海南花卉大世界"首届花卉节"、"海南首届国际盆景根艺赏石邀请展"暨"海南省第三届盆景评比展览"、"海南省第九届迎春花市"、海南花卉大世界 "第二届花卉节"暨"海南省第十届迎春花市"、"海南首届藏文化暨藏獒展"、"海南省第十一届迎春花市"、"2012年东盟国家旅游商品中国巡回展"等。

Hainan Flower World has received much concern and support of the government since its opening in 2009, playing an important role not only in the development of Hainan's industry of flowers and plants, but also the construction of international tourism island.

1. 海南花卉大世界开园暨首届迎春花卉节

笔者主持开园仪式并致辞

市委有关领导及商家优秀代表参加"海南花卉大世界项目推介会"后合影

2. 首届国际盆景邀请展

笔者主持海南花卉大世界首届国际盆景邀请展示范表演颁奖（中左为深圳张辉明先生、中右为马来西亚陈官顺先生）

3. 世界盆景友好联盟揭牌仪式

WBFF 胡运骅主席致辞

台湾女主持兼翻译和笔者共同主持仪式

笔者迎接国际贵宾

WBFF 世界盆景友好联盟16个国家负责人或代表参加仪式

4. 第九届迎春花市

左起：笔者、曹五院、余长龙、张其光、黄金城、李慰春、林劲、刘立武、连明良共同出席省九届迎春花市

笔者主持海南省第九届迎春花市

5. 第十届迎春花市

笔者接受海南电视台等单位采访

左起：笔者、吴云茂、孙道静、刘艳玲、林劲、沈处长、廖小平出席剪彩

6. 第十一届迎春花市

笔者和海南省台盟主委连介德（左3）、省林业局原局长朱选成（左4）、省林业局副局长刘艳玲（左5）、市政府副市长蒙国海（左6）、市林业局局长林劲（左8）、市园林局局长江长桥（左1）及张华江董事长（左7）共同出席开幕式

笔者在海南省第十一届迎春花市上致欢迎辞（2012年1月8日）

领导嘉宾剪彩

海南省第十一届迎春花市开幕式现场

7. 首届藏文化暨藏獒展

笔者在"海南花卉大世界首届藏文化暨藏獒展"开幕式上致辞

原西藏拉萨市市长洛卡（右4）、台盟海南省委主委连介德（右3）、副主委蔡扬生（左5）、海口市副市长蒙国海（右5）、市林业局局长林劲、市农业局局长张志坚（左2）、省委党校教授夏鲁平（左6）、西藏藏獒协会会长王永刚（左3）、公司副董事长陈颖昌（左4）、主持人陈爱京（左1）及笔者共同出席开幕式

刘传刚盆景艺术轨迹
Personal Record of Liu Chuangang

（一）国内部分

1982年5月　参加工作；

1983年10月　任湖北省咸宁地区花木盆景协会副秘书长、技术员；

1985年3月　任湖北省鄂南盆景园副主任、助理工程师；

1987年9月　任湖北省鄂南盆景园主任、工程师；

1991年7月　任咸宁地区盆景研究会副会长兼秘书长；

1993年5月　中国盆景艺术家协会第一届理事会推选为特邀理事；

1993年5月　创办海南中南盆景根艺联营公司任总经理兼法人代表；

1993年8月　任海南省盆景艺术家协会副会长；

1993年9月　中国盆景艺术家协会推选为第二届理事会理事；

1994年5月　任中国风景园林学会花卉盆景赏石分会委员；

1995年8月　任湖北省盆景研究会委员；

1998年1月　中国盆景艺术家协会推选为第三届理事会常务理事；

1999年8月　任海南省老年大学盆景讲师；

1999年11月　任澳门杯全国盆景精品展评委；

2000年6月　被聘为海南职业技术学院园艺系专业指导委员会委员；

2000年8月　被聘为海南职业技术学院园艺系兼职副教授；

2000年10月　任第十二期（湛江）中国盆景与园艺培训班授课老师；

2001年5月　任第十三期（成都）中国盆景与园艺培训班授课老师；

2001年5月　被聘为"蜀汉杯·中国盆景艺术作品展"评委；

2001年5月　被聘为海口经济技术学院客座副教授；

2001年6月　创办海南省花协盆景专业委员会，任主任；

2001年10月　主持海南省第一届盆景评比展览任总负责人；

2001年10月　被聘为《花木盆景》杂志编委；

2001年12月　被聘为《中国花卉盆景》杂志编委；

2002年4月　任中国风景园林学会花卉盆景赏石分会理事；

2002年10月　国家邮政部批准发行《刘传刚盆景艺术个性化邮票》一套16枚，面值80分；

2002年12月　在海口举行《刘传刚盆景艺术个性化邮票》首发式；

2002年12月　撰写的《博兰盆景的制作与养护》一文发表于《中国花卉盆景》2002年第12期上；

2003年10月　主持海南省第二届盆景评比展览任总负责人；

2004年1月　受聘为华南热带农业大学高职院园林系兼职教授；

2004年7月　受聘为海南省根雕文化艺术协会名誉会长；

2004年9月　受聘为海南职业技术学院生物科学系兼职教授；

2005年4月　受聘为海南省艺术家促进会"首席顾问"；

2005年5月　被中国盆景艺术家协会授予"中国盆景艺术大师"称号；

2005年9月　被聘为北京第八届亚太地区盆景赏石会议暨展览会评委；

2005年9月　被聘为第六届中国花卉博览会暨第四届精品盆景展评委；

2005年10月　被聘为首届中国（南京）绿化博览会盆景评委；

2006年4月　被聘为中国（陈村）BCI国际盆景赏石博览会盆景评委；

2007年5月　被聘为首届中国（陈村）盆景精品大奖赛专家评委；

2007年9月　被聘为第六届中国（厦门）国际园林博览会盆景评委；

2007年9月　撰写的《"雨林式"盆景的创作》一文发表于《花木盆景》2007年第9期上；

2008年9月　被聘为西安首届中国唐风展盆景评委；

2009年10月　被聘为广西盆景艺术大师评审委员会专家评委；

2009年12月　主持首届海南国际盆景赏石邀请展暨海南省第三届盆景评比展览任总负责人；

2010年1月　撰写的《弘扬岭南盆景，打造全新国际品牌》一文发表于《广东园林》2010年第1期上；

2010年6月　被聘为广州国际盆景邀请展评委及盆景高峰论坛主讲人之一；

2010年10月　被聘为上海"东沃杯"中国盆景创作比赛评委；

2010年10月　撰写的《"浅谈杂木类风动式盆景的制作"——以海南博兰树种为例》一文发表于《花木盆景（盆景赏石）》2010年第10期上；

2011年1月　赴香港参加"香港岭南盆景艺术学会第二届会员代表大会"、考察香港盆景并与港盆景界进行交流；

2011年1月　撰写的《盆景创作示范表演之我见》一文发表于《中国盆景赏石》2011年第1期上；

2011年11月　应台湾休闲农业组织之邀，赴台湾考察交流，其间拜访台湾著名盆景大师梁悦美教授。

2011年12月　被海口市林业局聘为市林业系统专家库专家；

2012年1月　被台盟海南省委会聘为参政议政委员会专家委员；

2012年5月　被海南师范大学聘为"特聘教授"；

2012年8月　撰写的《艺海启航》专著，在中国林业出版社出版。

（二）国际部分

2001年5月　应邀赴新加坡参加国际盆景赏石展并作盆景艺术演讲；

2003年2月　为中国盆景代表团成员在甘伟林理事长带领下赴菲律宾参加亚太盆展及申办第八届亚太盆景会议暨展览会获得成功；

2006年9月　应邀赴马来西亚参加国际盆景古石展赛会担任盆景评委及评审会主委；

2008年2月　为中国盆景代表团成员在胡运骅、辛常宝先生带领下参观日本"国风展"并在日本参观考察和交流；

2008年10月　BCI国际盆栽赏石协会授予"国际盆栽大师"称号；

2009年7月　为中国盆景代表团成员赴波多黎各参加世界盆景大会和参加申办WBFF世界盆景友好联盟落户中国获得成功；

后和黄就伟、林南一行三人赴美国西雅图等地进行盆景交流并作多场盆景创作示范表演；

2011年4月　为WBFF中国区成员在胡运骅主席带领下赴荷兰、西班牙、意大利进行盆景推广宣传和交流，其间在西班牙国际盆景展览会上做了两场风动式盆景创作示范表演；

小丛林（博兰）
Little Jungle(Poilaniella fragilis)

后 记
Postscript

我出生在湖北鄂州、沼山山麓、梁子湖畔的一个如水墨山水画中的村子里，祖上几辈务农。童年的我和小伙伴们，光着脚丫在田野里撒欢。记得那时节，我经常采摘一些不知名的野花，捡些漂亮的小石子，好奇一些奇奇怪怪的根兜，那里的山山水水滋润着我幼小的心灵。

1982年是我人生的第一个转折，我被分配到咸宁地区林业部门工作。当年，地区林业局成立盆景小组，时任林业局局长李日光先生有意栽培我，我便成为了其中一员，并随后成为这项工作的地区盆景园负责人。当我第一次看到盆景图片时，一刹那激发我寻找童年的挚爱，原来盆中也有这样美丽的世界！如是，如饥似渴地学习盆景艺术方面的知识，可是那年头盆景书籍太少太少。然而，我又太幸运了，1982年竟被贺淦荪先生收为弟子！在恩师的谆谆教诲下，我领悟到了做人的道理与盆景艺术之道。我一头扎进盆景创作之中，对恩师的一些作品，临摹一遍又一遍。为调查和采集湖北盆景资源，两次车祸中，大难不死……同时，也暗暗下定决心，把对学习、创作盆景点点滴滴体会记录下来，如果能对今后的自己或对别人有点点借鉴、点点益处，我就心满意足了。

1992年，随着时代的潮流，我来到了第二故乡海南。感谢我的妻子王四英同我一道，在这里相互扶持，历经千辛万苦、酸辣苦甜、尝尽人生百味。但在盆景艺术的探讨上，我有幸踏入神秘的热带植物天堂，以至引发我对海南博兰、香兰、黄花梨盆景以及雨林式盆景等创作的冲动，致使一批盆景系列作品问世。虽然对这些作品褒贬各异，可我认为：都是用不同的视角来看待，故而能得出不同的结论。对于来自各方面的不同意见，都是对我极大的鼓舞和鞭策，我会虚心、认真地整理、归纳、吸收，以至不断完善我的盆景创作。

伯父刘定一题字

笔者专赴台北拜访92岁高龄的旅台著名文化人、书法家、伯父刘定一先生（右）并在其家中留影

笔者全家福

    2007年，又是我一次人生的大转折，由于我创建的原"海口市刘传刚盆景艺术中心"面临开发商占地搞房地产，并被迫搬迁，海南行内数十家商户和我一样也临同运。鉴于此，作为行业会长有责任应该做点什么，故联合同仁侯立军先生等人于同年创建了"海南花卉大世界"项目，后又与著名企业家张华江、陈颖昌先生等联袂合作，使该项目不断走向成熟或完善，得到省、市、区各级政府和领导重视，现正为创建国家4A级旅游景区而努力。与此同时，我在花卉大世界园区内创办了自己的盆景园——锦园，倾注诸多心血，得到了各级领导、专家和同仁的关爱与支持，使自己盆艺事业得以延续和发展。

    艺海无涯。我才刚刚启航，无论是碧波万顷还是雨骤风狂，我深信：艺术的航船都会驶入一个美丽的港湾。

    I was born of a farmer's family in Hubei province. The village I lived in was filled with natural beauty, cultivating my heart and mind.

    In the year of 1982, I was assigned to work in the ministry of forestry in Xianning region and began to learn how to train works of penjing. Later, I formally acknowledged Mr. He Gansun as my master and enjoyed the great benefit of his instructions.

    In 1992, I left Hubei for Hainan Island with my family. Thanks to the support of my wife, I reached a new stage by creating works with local trees in Hainan.

    In 2007, I met a turn in the course of my career with the project of Hainan Flower World, which can be regarded as a new starting point for me.

    There is no end in the ocean of art, and I just begin to sail.

刘传刚

2012年5月1日于海口

鸣 谢
Special Thanks

◎ 中国盆景艺术大师、恩师贺淦荪先生
　　　　　　　　　　　　　　—— 为本书题写书名

◎ WBFF 世界盆景友好联盟主席胡运骅先生
　　　　　　　　　　　　　　—— 为本书作序

◎ 中国盆景艺术大师胡乐国先生
　中国盆景艺术大师赵庆泉先生
　　　　　　　　　　　　　　—— 为本书点评

◎ 全国政协副主席、台盟中央主席林文漪女士
　全国政协人口资源环境委员会副主任、中共海南省委原书记汪啸风先生
　海南省政协副主席、中国书法家协会理事王应际先生
　中国当代著名学者、国学大师、教授范曾先生
　中国盆景艺术家协会永久名誉会长苏本一先生
　旅台著名文化人、书法家、伯父刘定一先生
　　　　　　　　　　　　　　—— 为本书题字

情洒琼州（三角梅，垂枝式造型）

◎ 以及四川吴敏先生顾问，中国林业出版社徐小英编审策划、赵芳女士设计，中国盆景高级艺术师唐吉青先生绘图，海南省高校摄影协会会长林伟智院长拍摄照片，海南《海风》杂志社陈水雄主编、海口经济学院刘荆洪教授及海南胡庆魁副教授作指导，陈依菁女士翻译等。

◎
- The name of this book is inscribed by Mr. He Gansun.
- This book is prefaced by Mr. Hu Yunhua.
- This book is inscribed by Mrs. Lin Wenyi, Mr. Wang Xiaofeng, Mr. Wang Yingji, Mr. Fan Zeng, Mr. Su Benyi and Mr. Liu Dingyi.
- Comments are given by Mr. Hu Yueguo and Mr. Zhao Qingquan.
- Producer: Xu Xiaoying
- Art Editor: Zhao Fang
- Illustrator: Tang Jiqing
- Photographer: Lin Weizhi
- Advisors: Wu Min, Chen Shuixiong
- Translator: Chen Yijing

南丽湖晨曦（榆树、六月雪、海石）
Daybreak(Ulmus Pumila, Serissa Japonica, Sea stone)

图书在版编目（CIP）数据

艺海启航：刘传刚盆景艺术轨迹／刘传刚著．－北京：中国林业出版社，2012.8
ISBN 978-7-5038-6706-4

Ⅰ.①艺…　Ⅱ.①刘…　Ⅲ.①盆景－观赏园艺　Ⅳ.①S688.1

中国版本图书馆CIP数据核字（2012）第182814号

出　版	中国林业出版社（100009　北京西城区刘海胡同7号） http://lycb.forestry.gov.cn E-mail:forestbook@163.com　电话：(010)83222880
发　行	中国林业出版社
制　版	北京捷艺轩彩印制版技术有限公司
印　刷	北京中科印刷有限公司
版　次	2012年8月第1版
印　次	2012年8月第1次
开　本	215mm×280mm
字　数	389千字
照　片	约467幅
印　张	11
印　数	1～3 000册
定　价	260.00元

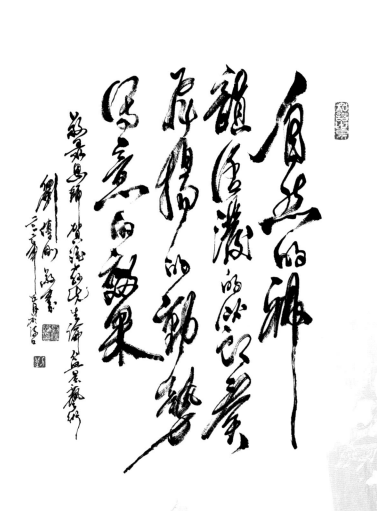